Principles of Petroleum Development Geology

Prentice Hall Petroleum Engineering Series

Larry W. Lake, Editor

Economides, Hill, and Economides, *Petroleum Production Systems*
Lake, *Enhanced Oil Recovery*
Laudon, *Principles of Petroleum Development Geology*
Raghavan, *Well Test Analysis*
Rogers, *Coalbed Methane*

Other Titles of Interest

Carcoana, *Applied Enhanced Oil Recovery*
Craft, Hawkins, and Terry, *Applied Petroleum Reservoir Engineering*
Schecter, *Oil Well Stimulation*
Whittaker, *Mud Logging Handbook*

Principles of Petroleum Development Geology

Professor Robert C. Laudon
University of Missouri
Rolla, Missouri

PTR Prentice Hall
Upper Saddle River, New Jersey 07458

Library of Congress Cataloging-in-Publication Data

Laudon, Robert C.
Principles of petroleum development geology / Robert C. Laudon.
p. cm.
Includes bibliographical references and index.
ISBN 0-13-649468-4
1. Petroleum—Geology. 2. Petroleum reserves. I. Title.
TN870.5L29 1996 95-24201
553.2'8--dc20 CIP

Editorial/production supervision: *bookworks*
Cover design: *Amy Rosen*
Cover photo: *Ken Condon, The Stockroom Illustration Source, Inc.*
Manufacturing manager: *Alexis Heydt*
Acquisitions editor: *Bernard Goodwin*
Cover director: *Jerry Votta*

Prentice-Hall, Inc.
A Pearson Education Company
Upper Saddle River, New Jersey 07458

Printed in the United States of America

ISBN 0-13-649468-4

Prentice-Hall International (UK) Limited, London
Prentice-Hall of Australia Pty. Limited, Sydney
Prentice-Hall Canada Inc., Toronto
Prentice-Hall Hispanoamericana, S.A., Mexico
Prentice-Hall of India Private Limited, New Delhi
Prentice-Hall of Japan, Inc., Tokyo
Pearson Education Asia Pte. Ltd., Singapore
Editora Prentice-Hall do Brasil, Ltda., Rio de Janeiro

This book is dedicated to my wife, Linda,
who has stayed with me and encouraged
me throughout the writing of this book.

Contents

Preface

The purpose of this book is to outline the fundamental principles of petroleum development geology. The book is not intended to be exhaustive, but rather an overview of the most fundamental concepts in petroleum geology. These concepts are important for exploration and environmental geologists as well as development geologists.

Petroleum development geology is a hybrid discipline. It is geology at the field and reservoir scale. It therefore contains strong elements of all of the following disciplines:

- Structural geology
- Stratigraphy and sedimentation
- Reservoir engineering
- Drilling engineering
- Petrophysics (well-log analysis)
- Reflection seismology
- Petroleum land management
- Economics
- Organic chemistry.

Each of these disciplines can be examined in detail in other textbooks that are excellent and exhaustive. This book is designed to outline the most salient aspects of these disciplines as they apply to development geology.

The book was originally written as course notes for a combination of courses taught at the University of Missouri, Rolla and a professional short course and is written on an introductory level. The text concentrates on principles, while practical field examples and problems appear in laboratory exercises.

Most textbooks on petroleum geology are organized in the sequence that oil takes in migrating from source rocks to traps:

1. Source rocks: origin of hydrocarbons
2. Reservoir rocks: characteristics of the reservoir
3. Traps: how oil and gas accumulate in the subsurface.

The structure of this book is different because the principal charge of the development geologist is the *estimation of volumetric reserves*. The principles necessary for the estimation of reserves—traps, contour maps, isopach maps, and volumetrics—are introduced at the very beginning in Chapters 1, 2, and 3.

Following a discussion of traps and volumetric reserve estimates, Chapters 4 and 5 deal with basic data—where it comes from, what it looks like, and its reliability. These chapters introduce some of the most fundamental concepts—well-log analysis, mud logging, plus core and cuttings descriptions. Facies analyses from core descriptions, but not direct interpretations, are introduced in later chapters (Chapters 12 and 13).

The section on traps and volumetrics is followed by a discussion of reservoir mechanics (Chapters 6 through 9). These chapters introduce fundamental concepts of pressure, temperature, porosity, permeability, relative permeability, well spacing, drive mechanisms, recovery efficiencies, and the chemistry of reservoir fluids, including formation volume factors. All of these are fundamental to the understanding of reservoir behavior and the estimation of volumetric reserves. Chapter 10 is a summary chapter on volumetrics, and discusses some cumulative errors that can occur.

Chapter 11 is one of the most important chapters in the book, and is unique among geology textbooks, as it introduces decline curves and engineering economics. These two subjects are combined because the mathematics of present value calculations are almost identical to those of exponential decline curves. These two concepts, used in tandem on a spreadsheet, provide a powerful means for creating full economic justification for individual wells or for whole field development projects. This chapter is also organized such that the most important equations are summarized on single pages that can be reproduced and used as "cheat sheets" in the office or in the field.

Chapters 12 and 13 introduce clastic and carbonate depositional systems. Concepts from these chapters are fundamental to reservoir characterization, reservoir modelling, and reservoir simulation on the computer.

Chapters 14 and 15 are the wave of the future for the development geologist. Chapter 14 introduces reflection seismology. The single most important new tool in the petroleum industry is three-dimensional seismic analysis, and one of the most important sources of additional domestic reserves in the future is EOR (enhanced oil recovery) discussed in Chapter 15.

Finally, Chapter 16 is a simple summary outlining the role of the computer in current and future development geology.

The concept of the development geologist is not universally accepted by all petroleum companies. In some companies, the exploration geologist serves also as a development geologist. In other companies there are no development geologists at all. Engineers cover all aspects of development design and drilling.

Development geology has grown through the years because managers have come to understand that engineering, geology, and geophysics are not just technically diverse disciplines, but that engineers, geologists, and geophysicists have fundamentally different thinking styles. Geology is simply not as precise as engineering and geophysics, and the artistic aspects and imprecision of geology are often difficult for engineers and geophysicists to appreciate. A communication problem exists among these disciplines, not because of the technical aspects, but primarily because of practitioners' differences in thinking styles and associated personalities. The field of development geology has grown because managers have come to recognize that thinking styles are as important as expertise for full and accurate analyses in a complicated and diverse business. But most importantly, the field has grown because development geologists can bridge the communication gap that has always existed among these diverse disciplines.

Development geology will thrive in the future because another energy crisis is coming. The United States has a very large, looming energy problem. The United States is in a mature state of energy development. The elephants (giant oil and gas fields) have been found. The size of discoveries has decreased markedly through time and will continue to decrease. Seventy-one percent of the world's oil and gas wells are in the United States, and they constitute less than 4 percent of the world's remaining reserves. These reserves are all declining and are not being supplemented by new reserves. As of this writing, the active domestic rig count is approximately 20 percent of the 1980 level. As of this writing, the United States has an imbalance of payments of approximately $200 per U.S. citizen per year for oil alone. Alternate energy sources will certainly continue to become more important in the future, but without subsidies, most are simply not feasible until the price of oil is at least $40 to $50 per barrel.

Development geology will become increasingly important because future domestic oil and gas reserve additions will come more from new opportunities in old areas, and less from traditional forms of rank wildcat exploration. In addition, development geology will support extracting additional reserves from existing reservoirs. Primary and secondary oil recovery techniques recover less than 40 percent of the original oil in place from most reservoirs around the world, leaving over 60 percent in the ground. This important natural resource will be exploited only through advances in our understanding of geology and engineering at the reservoir scale. This is the domain of the development geologist.

Although this book is entitled *Principles of Petroleum Development Geology,* many of the principles introduced are important to both exploration geology and environmental geology. The principles governing how nonaqueous phase contaminant fluids (DNAPLs and RNAPLs) flow and are trapped in the subsurface are almost identical to the principles dictating how oil and gas flow and are trapped in the subsurface. Exploration geologists make reserve estimates every day, but in many cases, they do not understand the details of the geology at the reservoir level.

When, not if, the next energy crisis occurs, this book will be invaluable to the next generation of petroleum geologists. These geologists are likely to be very good with hi-tech gizmos, but will be very inexperienced with the details of how to actually find oil. Many will never have been involved with the actual drilling of a well, and though it has been said that "the largest amount of oil has been found in the mind of man," the oil is not proved until a well goes down or a recovery project is operational. The next generation of petroleum geologists will be making the largest reserve additions not by rank wildcat exploration, but rather by recognizing new opportunities in old areas.

Acknowledgments

This book is a compendium of ideas gained over the years from work at Shell Oil Company, DeGolyer and MacNaughton, and 15 years of teaching at the University of Missouri, Rolla. While it is not possible to thank everyone for everything, I would like to single out a few people whose experience and knowledge has contributed greatly to my understanding of the petroleum industry.

From DeGolyer and MacNaughton, I wish to acknowledge the help of Fred Grote, Dick Elliot, and John Olson. From Shell, I express my appreciation to Jim Hartman, Chuck Speice, Bruce Bernard, Dan Stevens, Howard Johnson, and the whole Shell training program. At the University of Texas, I wish to thank Bill Fisher, Frank Brown, and Alan Scott for unifying the science of sedimentary rocks in the concept of depositional systems. I also wish to recognize Earle McBride and Bob Folk at the University of Texas. At the University of Missouri, Rolla, I wish to thank Jerry Rupert for input on geophysics. I also thank Len Koederitz for helping me understand pressure buildup and drawdown curves. I wish to thank International Human Resource Development Corporation (IHRDC) of Boston for the opportunity to teach a Development Geology short course whose course notes became part of this text. I also wish to thank those who suffered through the first teachings of that short course.

Lastly, I wish to thank Rich Green and my brother, Dick Laudon, who have been good friends and encouraged me throughout the writing of this book. When I needed information on current trends or independents in the industry, they were always there to answer questions.

Principles of Petroleum Development Geology

CHAPTER 1

Introduction

1.1 What Is Development Geology?

After an exploration group discovers a field, the geologic responsibilities for the field are normally turned over to a development geologist. The primary function of the development geologist is to develop the field as economically and efficiently as possible. This requires input from geophysicists, exploration geologists, economists, reservoir engineers, drilling engineers, petrophysicists (well-log analysts), and people from other engineering disciplines. The development geologist is in a central position to all of these disciplines (Fig. 1–1) and must be able to communicate effectively with all of these people.

The principal responsibilities of a development geologist can be classified into four broad areas:

1. **Predevelopment evaluation:** After a field has been discovered, exploratory wells and delineation wells are drilled to evaluate the field for reserves and design criteria. This phase is particularly important in offshore areas where large capital investments are involved and incorrect or imperfect designs can be extremely costly.
2. **Development drilling:** During the actual drilling of development wells the development geologist is responsible for:
   (a) Initiating development well recommendations
   (b) Monitoring these wells while they are drilled
   (c) Adjusting development plans as wells are drilled

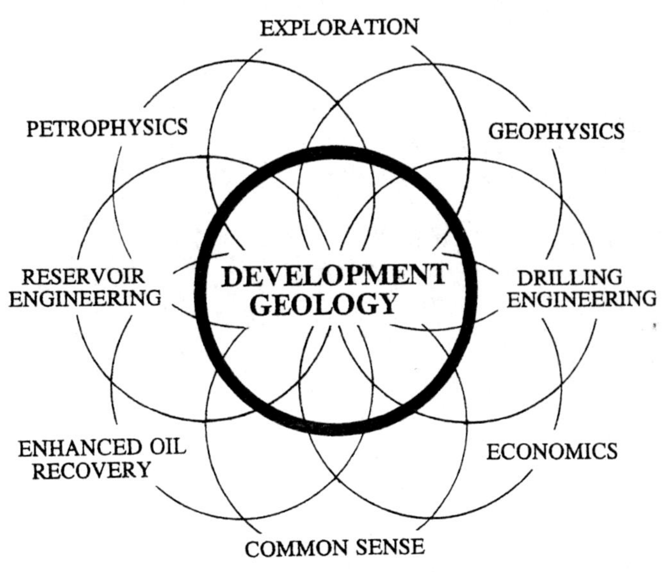

Figure 1.1
Development geology is a hyprid discipline whose strength is derived from communication between and coordination of all the surrounding groups. Ideally, all of the circles overlap somewhat.

3. **Well surveillance:** Most surveillance of on-line wells is handled by the reservoir engineer. But when reservoirs are not performing as expected or when remedial work on a well (workover) is required, geologic input can be extremely important. The reservoir engineer and geologist must work together to evaluate unusual reservoir performance and to make recommendations for remedial operations.

4. **Field studies:** One of the most important functions of the development geologist is to reevaluate and recognize new opportunities in old fields.

The fourth item will become more important in the future, especially in the United States, as domestic production declines and reserves are not replaced by conventional exploration methods. Future domestic reserve additions will be based more on the reevaluation and recognition of new opportunities in old fields than on conventional wildcat exploration.

1.2 Synergism

Professionals in the separate disciplines of geology, geophysics, and engineering have their own vocabularies and their own ways of thinking. The most important point of this book is synergism: the three disciplines' modes of thinking and vocabularies are important and, *collectively,* they are effective in the proper evaluation, development, and management of oil and gas reservoirs. Geologists, engineers, and geophysicists must work together to solve the complex problems of today's oil-exploration world.

Good development geologists have two very important characteristics.

1. They can communicate and integrate on all levels with geophysicists, engineers, and field personnel.
2. They can conceptualize the earth in three dimensions and can graphically represent those concepts on maps.

The first quality is primarily one of personality. Such an individual is typically personable, with a strong background in geology to appreciate the enormous complexities of the earth. At the same time this person must be able to communicate effectively with engineers and geophysicists on their own ground and to make geologic sense of their mathematical models.

The second characteristic is very difficult to teach. Not all people can see in three dimensions, and even for those who can, the earth's complex structures and stratigraphy are a constant challenge. The best development geologist can visualize several interpretations at once and be able to decide which is the better. Computers are very good at posting data and contouring simple structures, but no computer can match the human mind for synthesizing imprecise data and conceptualizing alternative hypotheses. Computers are a great help to the development geologist, but there is no substitute for experience when it comes to the mapping of geologically complex fields.

1.3 Development Geology in a Major Petroleum Company

Most vertically integrated petroleum companies recognize at least five general subdivisions:

1. Exploration
2. Production
3. Refining

4. Transportation
5. Marketing

Some companies combine exploration and production into one division (E & P), and others combine marketing, transportation, and refining in various combinations. Table 1–1 shows the major positions, their job descriptions, and normal disciplines hired to fill those positions within exploration and production departments of most petroleum companies. The list is not complete, but it shows that geology is important in both general subdivisions.

Exploration is the business of discovering commercial deposits of oil and gas. The business of the production department is to develop and produce the oil and gas deposits in the most efficient and economical means possible. In the past, most exploration has been the domain of the geologist and geophysicist, while production has been the domain of the petroleum engineer.

Most companies have found that geologic thinking is extremely important during the development phase for most oil fields, and the more complicated the field the more important the geological interpretation becomes. A development geologist has expertise in both exploration and engineering, and is generally a geologist with a strong engineering background, or less commonly, an engineer with a strong geology background.

1.4 Development Geology in Independent and Small Oil Companies

Small companies rarely have the resources to hire specialists in all of the specialty areas. Commonly a geologist/geophysicist combination constitutes the exploration department and a petroleum engineer constitutes the production department. In a truly independent company, one person constitutes all of E&P, and consultants are hired to fill in the expertise gaps.

One of the most exciting and potentially rewarding professions anywhere is that of independent petroleum geologist. In order to understand what makes the business so exciting, one must understand how petroleum deals are put together and the concept of an overriding royalty interest (ORI).

Table 1–2 shows how a prospect or idea gets financed by means of the famous (some say infamous) “third for a quarter” deal. Although the “third for a quarter” deal is only one of many possible financial arrangements, the concept is important because it illustrates how the prospect

Table 1–1 Common employment positions, employment descriptions, and disciplines normally hired to fill those positions in an exploration and production (E&P) department of a vertically integrated oil company.

Divisions	Positions	Job Description	Normal Disciplines Employed
Exploration	Seismic Field Crew	Runs seismic lines in field	Geophysicist, Mechanic
	Seismic Processor	Makes seismic lines look good	Geophysicist, Mathematician, Computer Programmer
	Seismic Interpreter	Makes geologic maps from seismic lines	Geophysicist, Geologist
	Prospect Generator	Creates oil and gas exploration opportunities	Geologist, Geophysicist
	Operations	Drills exploratory wells	Geologist, Petroleum Engineer
	Field Geologist	Sits on wells; made surface geologic maps in past	Geologist, Paleontologist
	Well-Log Analyst	Makes certain logs are run properly and interprets logs	Geologist, Petroleum Engineer, Geological Engineer
	Landman	Buys and sells leases	Real Estate, Attorney
Production	Development Geologist	Production Geologist or Geological Engineer	Geologist, Geological Engineer
	Well-Log Analyst	Makes certain logs are run properly and interprets logs	Geologist, Petroleum Engineer, Geological Engineer
	Reservoir Engineer	Monitors reservoirs and makes economic projections	Petroleum Engineer, Geologist
	Drilling Engineer	Drills wells	Petroleum Engineer
	Production Engineer	Monitors reservoir production and makes workover recommendations	Petroleum Engineer
	Construction	Design platforms, pipelines, and production facilities	Civil Engineer, Mechanical Engineer

Table 1–2 Common structure for petroleum "deal" identifying how a petroleum geologist can obtain an overriding royalty interest in a well and how his/her company can get a beneficial interest in the well.

People	Financial Obligation	Financial Return on Investment
1. Landowner		
(a) Farmer		
(b) Mineral rights owner	None	Commonly 1/8 or 12.5% but negotiable
(c) Federal government		
(d) State government		
2. Promoters	None	Company gets 1/8 or 12.5% Prospect generator gets commonly 1 to 5%
(a) Prospect generator (usually geologist)		
(b) Intermediary (commonly landman)		
3. Working Interest Owners		
(a) Operator (usually the principal interest owner)	1/3 (33%)	1/4* (25%)*
(b) Other investors	2/3 (67%)	2/4* or 50*
		* Subject to above royalty covenants if other than 1/8
Totals >>>>>	Pays for 100% of the well	Accounts for all of whatever comes out of the well

generator may obtain an overriding royalty interest and his company may obtain a beneficial interest.

In the "third for a quarter" deal, some combination of investors must be found who will agree to finance 100 percent of the cost of a well in return for 75 percent of the income that results from the well. For example, if three investors each agree to purchase the "third for a quarter" deal, the three investors will pay 3/3 (100 percent) of the cost of the well, but they will own only 3/4 (75 percent) of any potential income that results from the well. This 75 percent interest is referred to as the *working interest,* and usually the selling company retains the remaining or *carried* 25 percent of the well.

In most areas of the U. S., some part of the revenue, commonly 1/8 (12.5 percent), must be paid to the landowner or mineral rights owner, an amount termed the *royalty interest.* Of the remaining 1/8 (12.5 percent), the company gets part (as an example, 9.5 percent) and the people who construct the deal get part (3 percent in our example). Commonly, 2 to 10 percent is available to the people who construct the deal. This final percentage is referred to as the *overriding royalty interest* (ORI). The people who construct the deal usually include the independent geologist(s) who generates the prospect and a landman who writes the contract with the mineral rights owner so that the prospect can be drilled.

An important concept is that the prospect generator, the independent geologist, can receive an overriding royalty interest purely for the idea of defining and justifying where a well should be drilled. The prospect generator need not invest his/her own money, and their company need not invest as much money as the working interest owners.

One does not need a background in calculus to figure out what this is potentially worth. If you, as a prospect generator, have a 3 percent ORI in a well that comes in at 500 barrels/day, you get 3 percent of the oil *off the top*. If oil is selling at $25 per barrel you will be making 3 percent of 500 barrels per day times $25 per barrel, or $375 per day, seven days a week, or over $11,000 per month from one well alone. In addition, if a discovery is made, the prospect generator normally receives the same ORI from all additional wells drilled on the lease.

Be certain to notice the difference between an overriding royalty interest (ORI) and a working interest. An ORI owner gets a percentage of the income off the top, before lease operating expenses, and has no financial obligations or liability. A working interest owner, an investor, receives income, but also has financial obligations for the drilling of the well, for operation of the producing well, and liability. If unusual expenses, such as a blowout, are incurred, the working interest owner has an obligation to pay his share of extra costs incurred.

The *operator* is the individual or company responsible for getting the well drilled. The operating company is usually the principal working interest owner in the field.

One can obtain an ORI from a patent, from writing a book, from mineral leases, or from generating prospects in the petroleum industry. Any royalty interest can be lucrative, but few are as lucrative as those from oil wells. It is the concept of the overriding royalty that makes petroleum geology unique, and this feature makes the average salary of petroleum geologists among the highest of any profession in the world today.

Most independent geologists are located in Houston, Dallas, Denver, New Orleans, Tulsa, Amarillo, Los Angeles, and other major oil centers. Most people in the U. S. know very little about these geologists, who are generally not in the business for the money, even though it can be very lucrative. Most enjoy the business of exploration; there is something very exciting about generating an idea or prospect, selling the prospect to a group of investors, watching the well go down day by day, and then explaining to the investors why it did or did not work. It is not for the faint of heart, but it does offer a rarely found form of job satisfaction.

1.5 Conclusions

How does a development geologist fit into this? Conventional exploration is in its old age in the United States. Conventional seismic lines are still run, and details of basins are still analyzed, but the domestic giant fields, commonly referred to as "elephants," and easy traps have all(?) been found. One or two domestic elephants may still be found, but they will never again be found at the rate that they were discovered during the 1920s through the 1950s. Future oil and gas, not just in the U. S. but also worldwide, is going to be much more difficult to find, and it is going to be found in smaller and smaller quantities by detailed geologic work. Through the reanalysis of older areas the development geologist is going to become the main explorationist within the domestic U. S. over the next 20 years, and throughout the world over the next 50 years. The development geologist has some powerful new tools—three-dimensional seismic, enhanced oil recovery (EOR) techniques, and horizontal drilling—that are going to affect the world's distribution of energy.

Finally, the bottom line on most major decisions is economics. It doesn't matter how good the geology or the engineering are, if the venture doesn't make economic sense, it won't get done. Proper economic evaluations require input from many different areas, and if the development geologist is to be included in the decision-making process, he/she must be able to understand and communicate effectively with people from a broad spectrum of engineering and scientific backgrounds as well as have a sound background in basic economics.

CHAPTER 2

Traps

In the evaluation of a petroleum exploration prospect, an exploration geologist will normally look for three essential components: a *source rock*, a *reservoir rock*, and a *trap*. Of these, by far the most important is the trap. Most geologists who have worked for a major oil company in a frontier situation can tell at least one story of justifying a rank wildcat well on the basis of the size of the anticline, never mind that the reservoir rock is Triassic schist or some other very unlikely reservoir rock, or that there is no identifiable source rock. If the trap is big enough, many major oil companies will drill it simply because of the large reserve potential.

Recent textbooks use a number of trap classifications. However, the vocabulary used in the petroleum industry will be used in this book and is outlined in Table 2–1.

Traps are composed of three important features:

1. **Reservoir rock:** A rock with enough porosity to be capable of storing economic quantities of petroleum, and enough permeability, either natural or induced, to be able to transmit the petroleum to a wellbore. The most common reservoir rocks are sandstone, limestone, and dolomite.
2. **Seal:** A relatively impermeable rock which disallows or retards the escape of petroleum moving upward through rocks.

Table 2–1 Summary of hydrocarbon trap types.

- I. Structural Traps
  - A. Dome (anticline with closure)
  - B. Anticline (closed by fault or permeability barrier)
  - C. Faults
    - 1. Normal faults
      - (a) Generic normal fault
      - (b) Growth faults
    - 2. Reverse faults
      - (a) Generic reverse fault
      - (b) Thrust faults
        - (i) Overthrust faults
        - (ii) Upthrusts
    - 3. Strike-slip faults
    - 4. Other fault types
      - (a) Scissors faults
      - (b) Transform faults
- II. Stratigraphic Traps
  - A. Primary stratigraphic traps
    - 1. Long, linear sandstone bodies
      - (a) Point bar sequences
      - (b) Delta distributary channel sandstones
      - (c) Barrier island sandstones
      - (d) Shelf sandstone bodies
      - (e) Submarine fan channels
      - (f) Natural pinch outs against topographic highs
    - 2. Natural facies changes
      - (a) Delta front sandstones
      - (b) Crevasse splays
      - (c) Turbidites
      - (d) Carbonate reefs
      - (e) Carbonate ramps
      - (f) Carbonate slopes
  - B. Secondary stratigraphic traps
    - 1. Angular unconformities
    - 2. Clay-filled channels (clay plugs, etc.)
    - 3. Diagenesis
      - (a) Cements
      - (b) Recrystallization
        - (i) Dolomite
        - (ii) Calcite
        - (iii) Chert
        - (iv) Clays
- III. Other Trap Types
  - A. Hydrodynamic
  - B. Unconventional (no definable trap, often source rocks)
    - 1. Fractures shales
    - 2. Some chalks

3. **Three-dimensional closure:** Some sort of three-dimensionally closing barrier against the base of the seal.

2.1 Anticlinal Theory of Traps

It was recognized very early that oil and gas tend to accumulate in structurally high areas. Early explorers mapped structures at the surface and projected those structures into the subsurface. It became very clear that gas and oil normally accumulate at the top of anticlines, as shown in Figure 2–1.

As source rocks, such as organic rich shales and evaporites in the subsurface, become compacted, they expel fluids, including water, gas, and oil. All of these fluids escape into available pore spaces, and as they migrate through any available permeability, much escapes to the surface. The petroleum becomes trapped in locations where water can continue onto the surface, but oil and/or gas cannot. In most cases the oil and gas are considered to have been trapped in the crest of the anticline by an almost totally impermeable seal, such as a shale or evaporite, in much the same manner as upward travelling gas bubbles would be trapped in an upside-down bowl in a bathtub full of water. Today many people feel

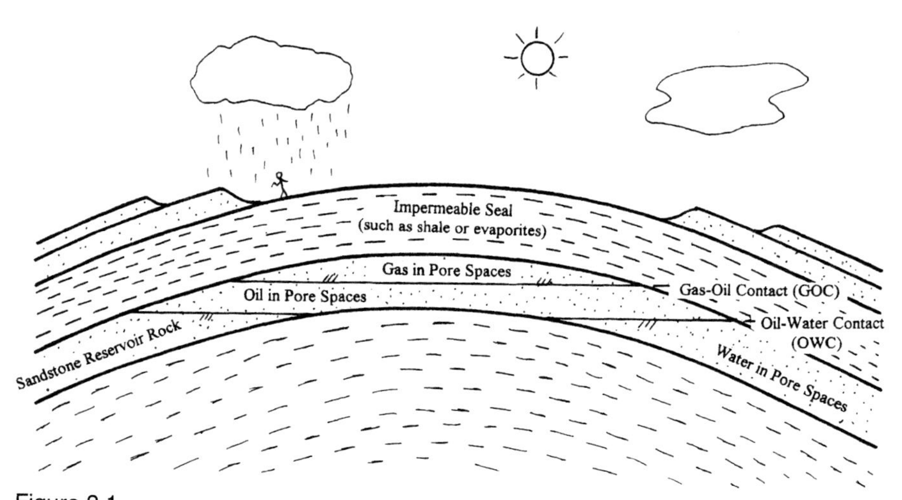

Figure 2.1
Cross section through a typical anticlinal trap showing a seal at the top, reservoir rock in the middle, and nonreservoir rock below. From top to bottom, the reservoir rock contains a gas cap, a gas-oil contact (GOC), an oil column, an oil-water contact (OWC), and water in the pore spaces below the OWC.

that the seal is not totally impermeable, but rather it acts as a semipermeable membrane allowing water, but not oil and gas, to pass through. Regardless of the permeability of the seal, oil and gas are trapped at the crest of an anticline (Fig. 2–1), and water is displaced.

The actual process of migration is considerably more complicated than the "upside-down bowl in a bathtub" analogy, which was, for the first 50 years of oil exploration, the main concept behind most oil exploration. Oil was simply trapped at the top of the anticline; this became known as the *anticlinal theory.*

To the purist, an anticline is a rock fold that is convex upward. Figure 2–2 shows structure contours on a plunging anticline. Such an anticline will not trap hydrocarbons because there is no three-dimensional trap. In fact, the term *anticlinal theory* is a misnomer, because an anticline, by itself, does not necessarily trap hydrocarbons.

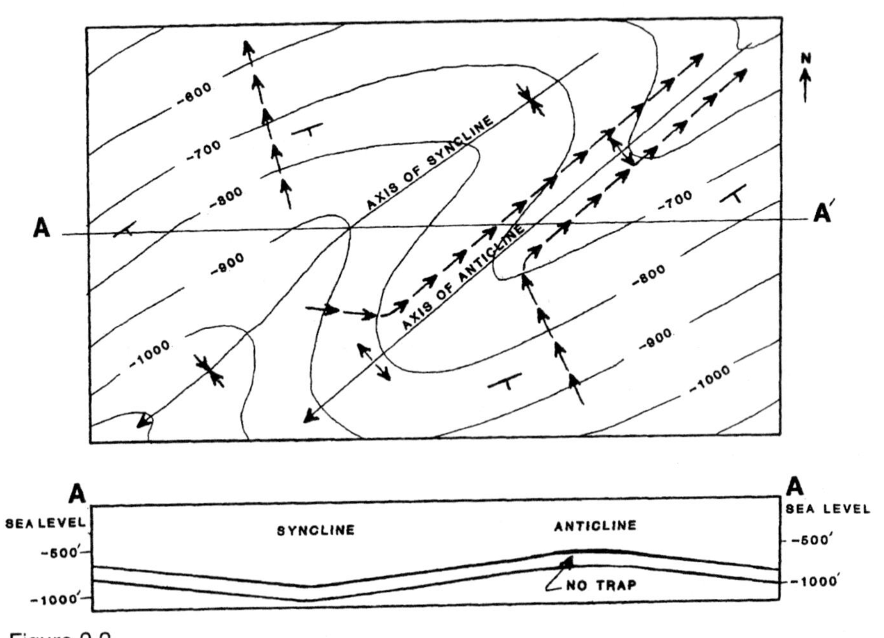

Figure 2.2
Map view and cross section of plunging syncline and non-closing anticline. This is an anticline with no trap. Arrows signify hydrocarbon paths of migration. Negative signs on structure contours indicate subsea (below sea level) elevations. Would the map be any different if the contour line elevations were positive?

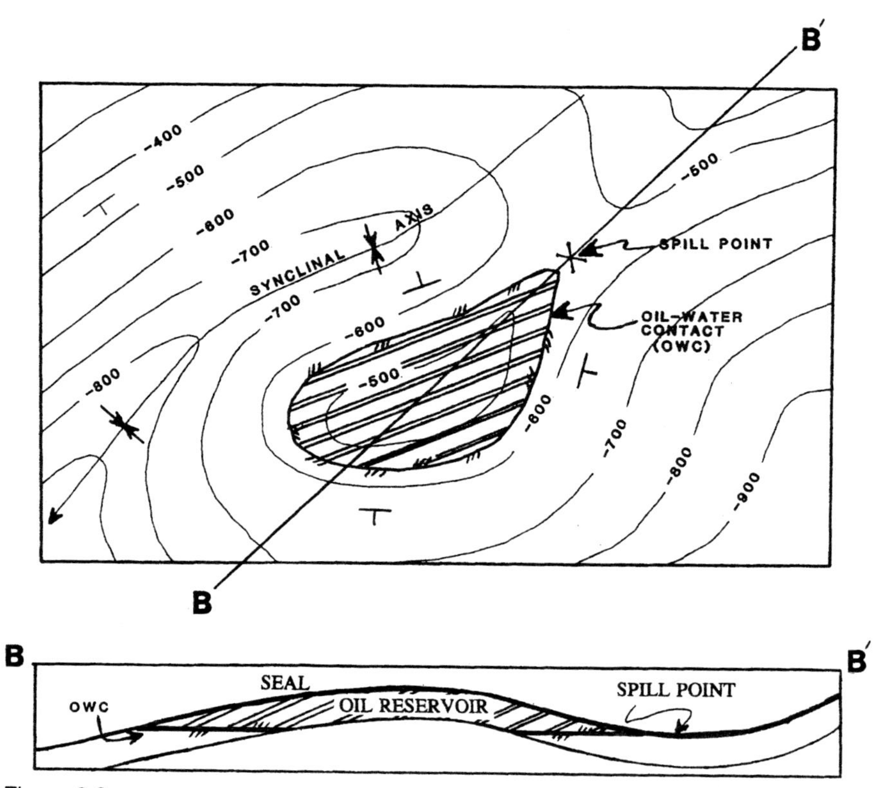

Figure 2.3
Structure contour map and cross section showing plunging syncline and anticline with closure, oil water contacts, and spill point.

In order for an anticline to trap hydrocarbons, the trap must be three-dimensional. That is, the anticline must be doubly plunging such that the structure contours close, or form a dome, as in Figure 2–3. Alternatively, the anticline may be truncated in some fashion, either by a fault as in Figure 2–4, or by a permeability barrier as in Figure 2–5, to form a *three-dimensional trap*. The anticlinal theory should really have been called the *domal theory,* because an anticline, by itself, will not necessarily trap oil.

The *spill point* occurs at the lowest point of closure on the trap and defines the maximum possible size of the trap. That is, as a trap fills with hydrocarbons it will continue to fill until the trap is either full to the spill point (Figs. 2–3, 2–4, 2–5) or until the trap starts to leak. At this

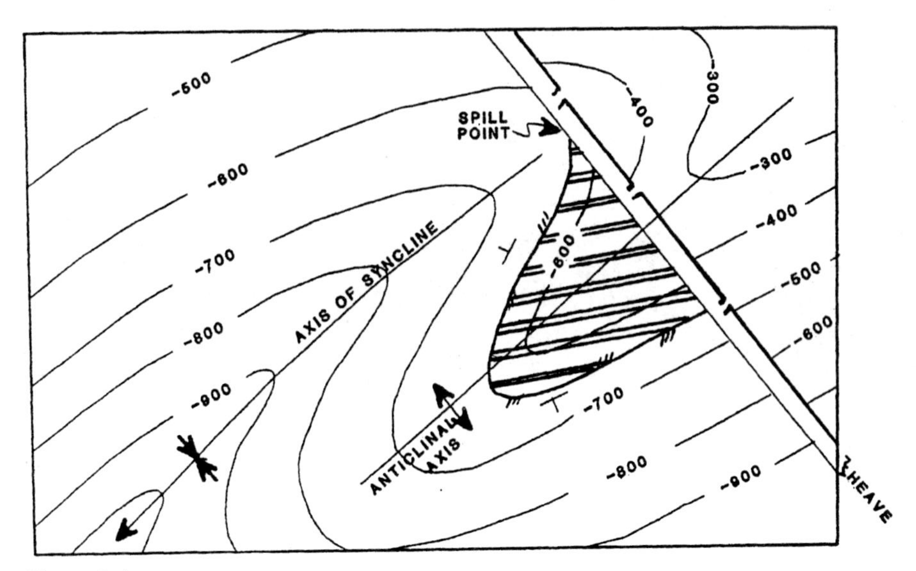

Figure 2.4
Plunging anticline, as in Figure 2.2, modified to form trap with closure completed by a sealing normal fault.

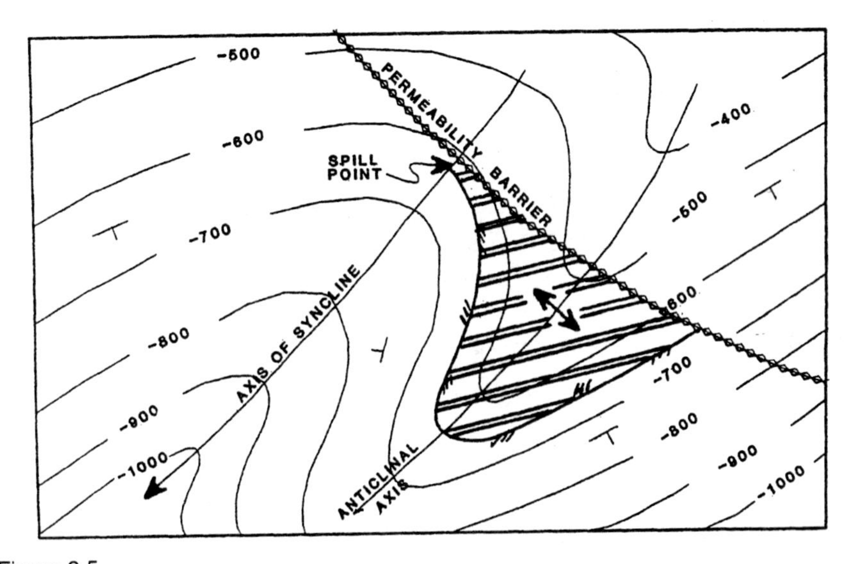

Figure 2.5
Plunging anticline, as in Figure 2.2, modified to form trap with closure completed by a permeability barrier.

point, hydrocarbons simply spill out of the trap and continue moving upward towards the surface.

2.2 Fault Traps

2.2.1 Normal faults

Figure 2–6 is a cross section through a garden-variety normal fault showing that the hanging wall is downthrown. Some relationships shown in this figure are quite important to the petroleum geologist and are not emphasized in most structural geology books. These points are:

1. **Normal faults are identified in vertical wells by the location of missing section.** *Caution:* Missing section can also occur at an unconformity, or could be the result of normal thinning. Also, a missing sand such as a missing channel sandstone, does not necessarily constitute missing section.
2. **The amount of missing section is the throw of the fault.** *Caution*: This is true only if the well is vertical. *Throw* is the vertical component of fault displacement. This is important because, in mapping, the difference between structure contours on opposite sides of a fault should match the throw of the fault. For example, if the contour on the upthrown side of a 800-foot fault is at –7000, then the structure contour line on the downthrown side should be –7800 (Fig. 2–7). For dipping rocks offset by a fault, vertical separation and throw may not be identical. Corrections for this error are not normally significant unless dips are steep.
3. **Heave is equal to the throw of the fault times the cotangent of the dip of the fault.** *Heave* is the horizontal displacement of a fault and is shown on a map as the width of the fault (Fig. 2–7). A well drilled into the heave area of a normal fault will not encounter the sand because it is faulted out at that location.
4. **Most normal faults dip at about 55 degrees in most field settings.** The cotangent of 55 degrees is 0.7, which means that the heave-to-throw ratio for most normal faults is 0.7. This means that 1,000-foot contour lines on a fault plane map should be about 700 feet apart. It also means that if the throw of a fault is 800 feet, then the fault separation (heave) on the map should be about 560 feet (Fig. 2–7).

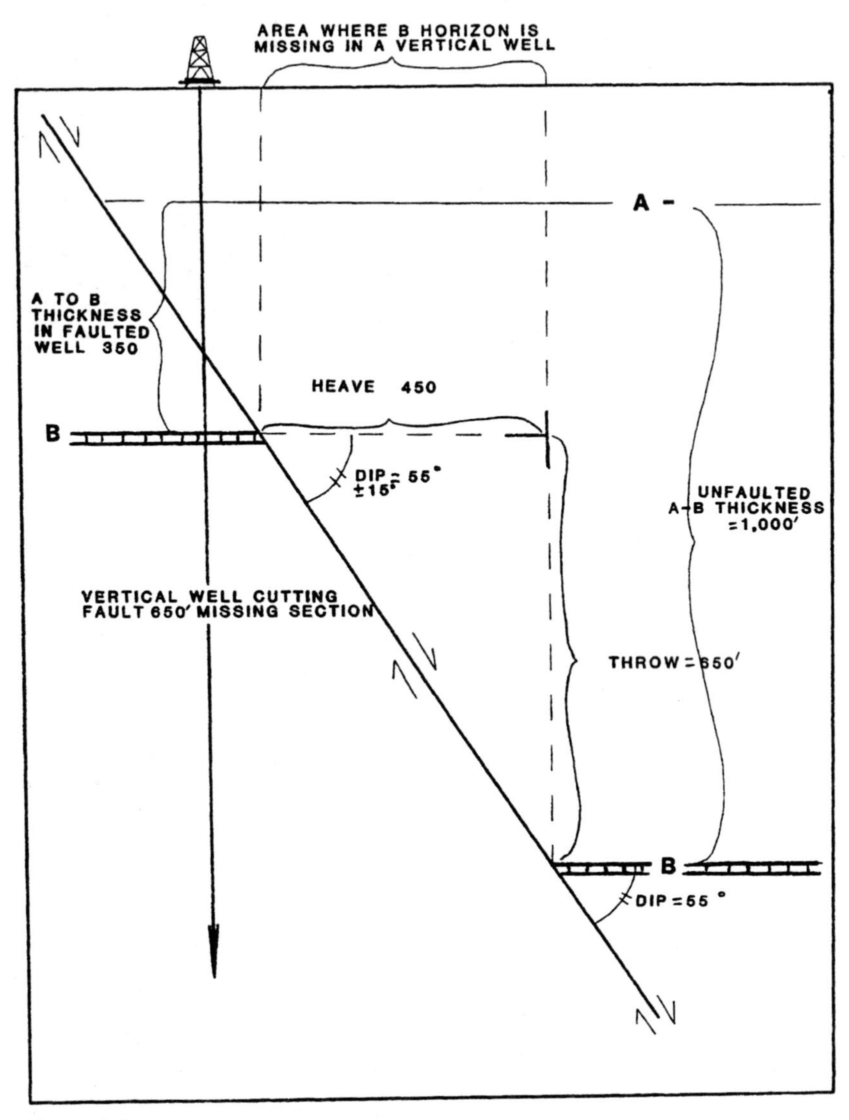

Figure 2.6
Cross-section view of a normal fault showing heave to throw relationship. *Throw* is the vertical component of fault displacement and is equal to the amount of vertical section missing in the log correlation. *Heave* is the horizontal component of fault displacement and is equal to the width of the gap on the strcuture contour map. A vertical well drilled in the heave zone will not intersect the mapped horizon.

5. **If these relationships are not honored on a structure contour map, the credibility of the contour map will be in doubt.** The importance of these rules is directly related to the scale of the geologic map. At exploration map scales of 1 inch =

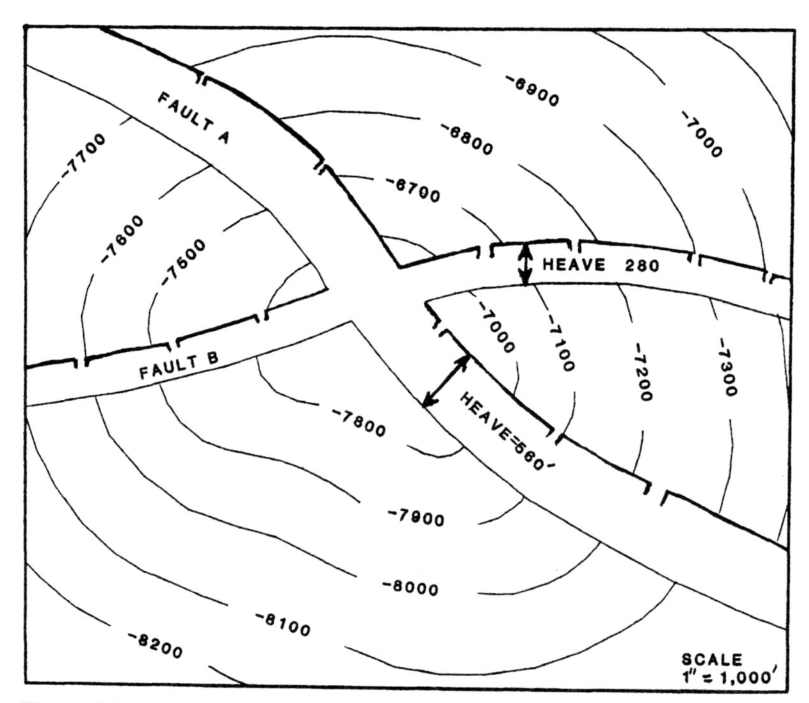

Figure 2.7
Structure contour map showing a normally faulted anticline. Notice that, by comparison of elevations on opposite sides of faults, throw on fault A is 800 feet, and throw on fault B is 400 feet. Assuming a fault dip of 55°, heave on fault A is 560 feet, or 800 * 0.7, and heave on fault B is 280 feet, or 400 * 0.7. At a scale of 1 inch = 1,000 feet, heave on fault A is 0.56 inches, and, heave on fault B is 0.28 inches. A vertical well drilled through the heave zone will not encounter the sand top. Tics on the fault always point down.

1 mile = 5280 feet or smaller, or in areas where faults are nearly vertical, these relationships are not important. But at field development scales of 1 inch = 1000 feet or greater, these relationships become quite important because accurate reserve estimates and proper placement of development wells are dependent on an accurate portrayal of the faults. Any geologist with Gulf Coast experience can tell almost instantly whether a particular map has been prepared carefully by checking these relationships.

2.2.2 Growth faults

A *listric normal* fault is a normal fault in which the hanging wall rotates, the fault plane itself is curved, and the dip of the fault decreases with depth. They are commonly associated with slumping and salt tectonics where salt has been evacuated from the toe of the fault.

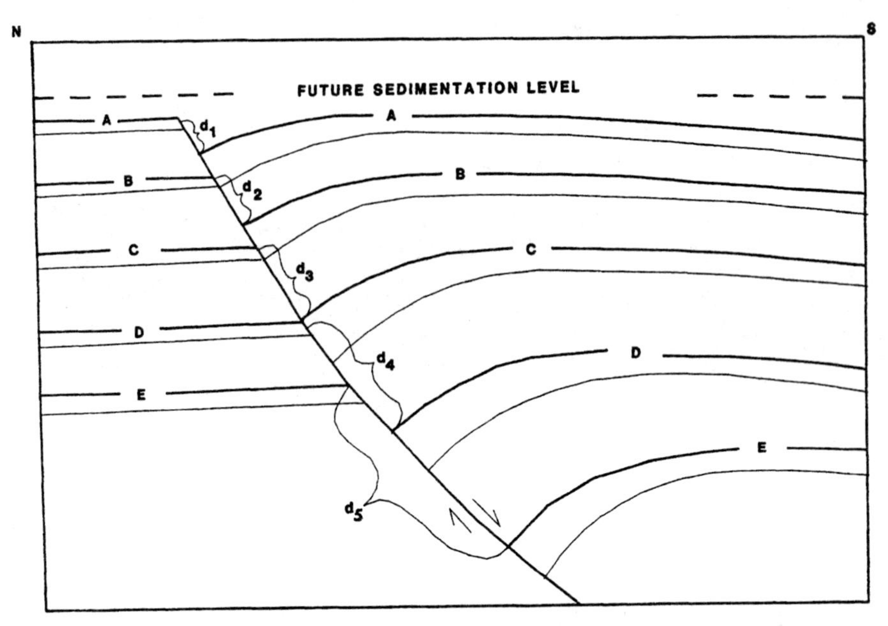

Figure 2.8
Cross section through a growth fault. Notice that the throw increases, or grows, with depth, and that the trap (crest of the anticline) is offset from the fault and migrates horizontally with depth.

A *growth fault* (Fig. 2–8) is the special case of a listric normal fault where sediment infilling at the surface has kept pace with fault movement such that the hanging wall has a considerably thicker sequence of sedimentary rocks than the footwall. Thus, the throw increases or grows with depth. Growth faults are commonly associated with major deltaic sedimentation, and are particularly common in the Gulf Coast where major deltaic sedimentation has occurred in association with salt tectonics.

Interestingly, the fault itself does not form the major trap. The most important trap is the rollover anticline, which occurs where the hanging wall rotates into the growth fault. The geometry of the rollover is that of reverse drag. The origin of this reverse drag is controversial, but undeniable. The crest of the rollover anticline may be several miles from the fault itself and, it is important to note, migrates with the fault. The crest approximately parallels the fault but is offset from the fault itself, and quite clearly a deviated hole must be drilled to penetrate the crest of the anticline continuously with depth.

Fields associated with growth faults are extremely important throughout the Texas and Louisiana Gulf Coast and Federal offshore waters. Excellent examples of such fields are illustrated in AAPG Memoir 14 (Tom O'Conner Field [Mills, 1974] and Vicksburg Fault Zone [Stanley, 1974]) and throughout the Transactions of the Gulf Coast Association of Geological Sciences volumes.

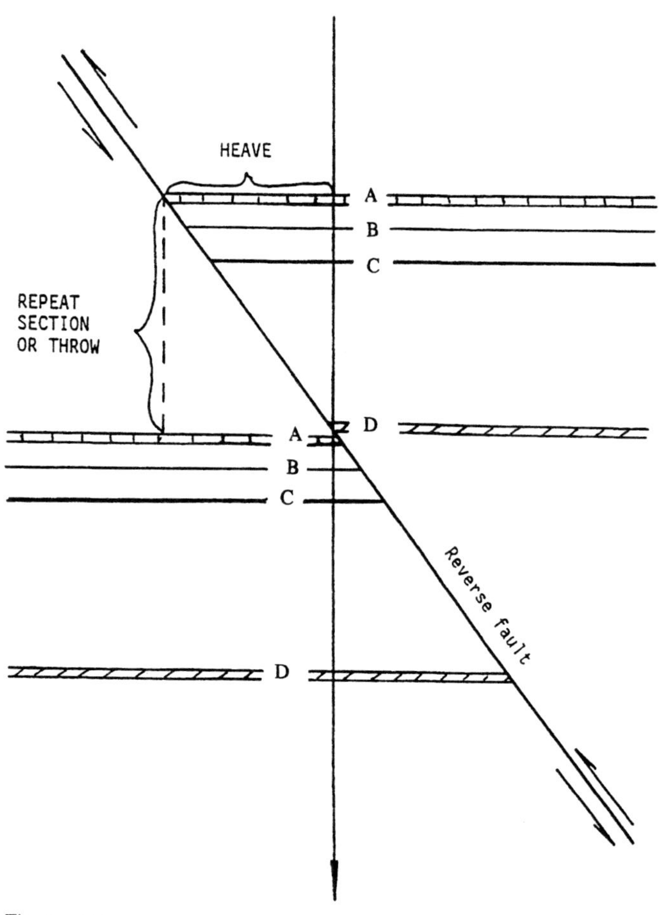

Figure 2.9
Cross section through a typical reverse fault. Note that a vertical well will encounter repeated section. Throw (vertical component) is equal to the amount of repeated section seen on logs. Heave (horizontal component) means that a well drilled in the heave zone will encounter the mapped horizon twice.

2.2.3 Reverse faults

Figure 2–9 (see previous page) is a cross section through a garden-variety reverse fault showing, by definition, that the hanging wall is upthrown. Several important points are implicit in the diagram:

1. A reverse fault is identified in a vertical well by the location of repeated section. *Caution:* In a highly deviated hole it is possible, but rare, to see repeated section on a normal fault.
2. The amount of repeated section is the throw.
3. Heave is equal to throw times the cotangent of the dip angle of the fault.
4. Thrust faults are low-angle reverse faults.
5. The dip angles on reverse faults, including thrust faults, are variable. Thus, the heave-to-throw relationships are not as consistent for reverse faults as for normal faults.

Thrust-faulted terrains tend to be very complicated, and a single diagram cannot represent all the trap possibilities. In North America the vast majority of petroleum associated with thrust fault terrains occurs in *leading-edge anticlines,* which are often recumbent meaning that one limb is overturned as in Fig. 2–10. In the fairway of the western overthrust belt of southwestern Wyoming and northeastern Utah, virtually all the discovered petroleum occurs in leading-edge anticlines similar to the Painter Reservoir (Fig. 2–11).

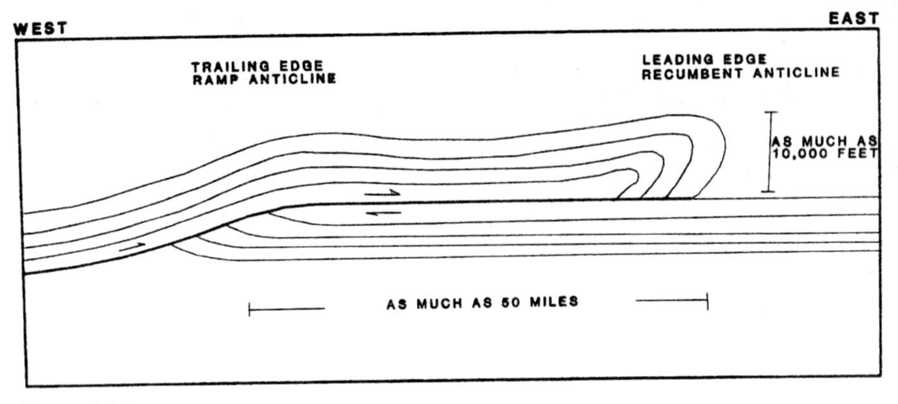

Figure 2.10
Schematic cross section through a thrust fault showing leading-edge recumbent anticline and trailing-edge anticline at the ramp. Most oil in the western overthrust belt of the U.S. is in the leading edge recumbent anticline.

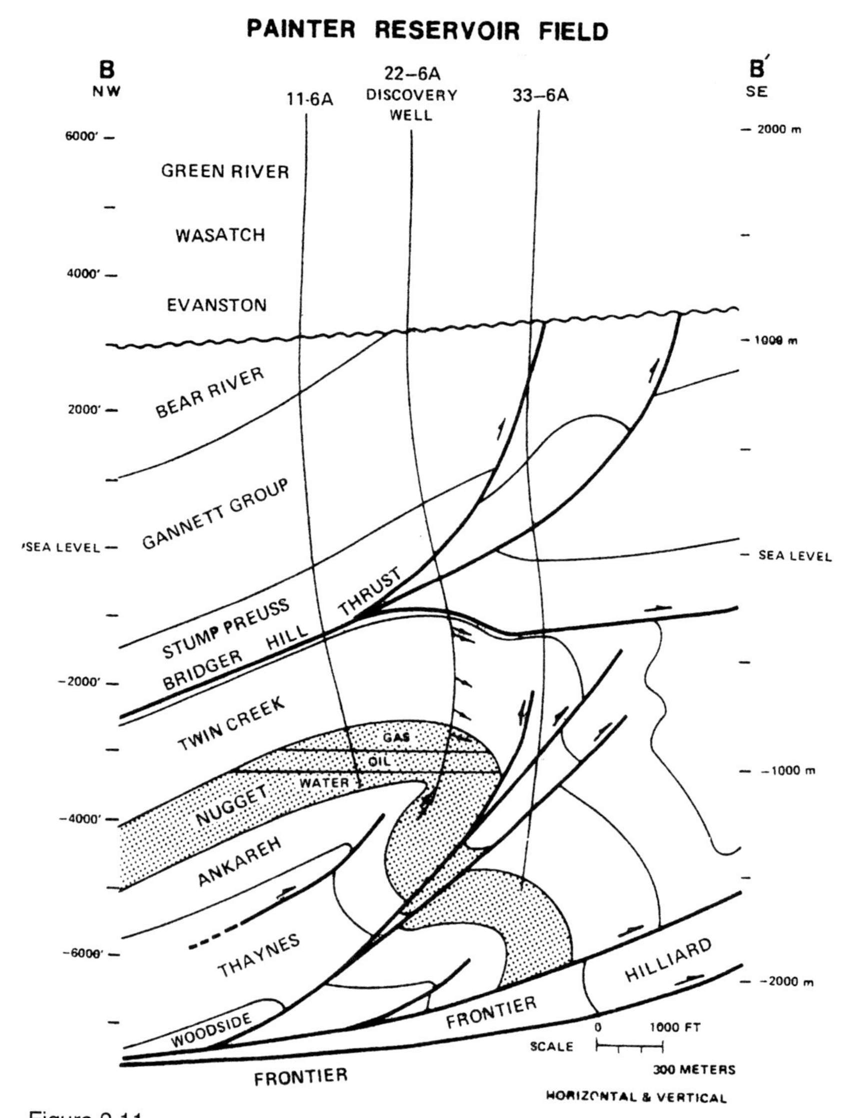

Figure 2.11
Cross section through the Painter Reservoir, a typical structure for the Wyoming Overthrust belt. From Lamb (180), reprinted by permission of AAPG.

Although leading edge anticlines contain the majority of oil found to date, other structures can also be important. Where a thrust fault ramps up over an underthrust plate (Fig. 2–10) an anticlinal structure in the overthrust plate is commonly formed. Other, more complex structures

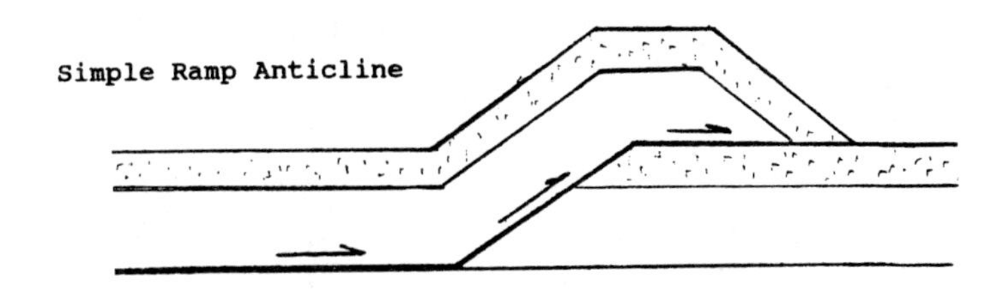

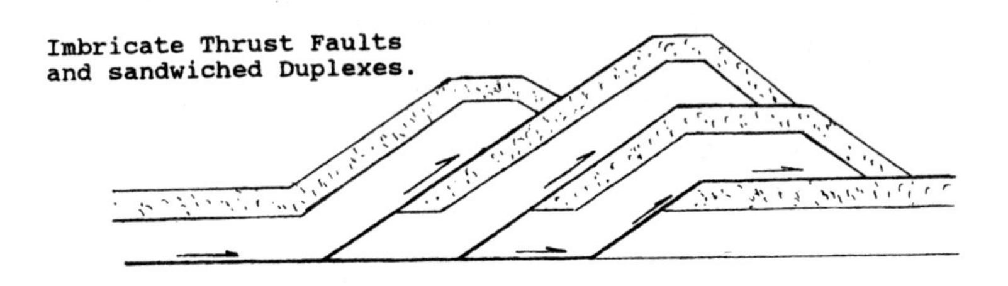

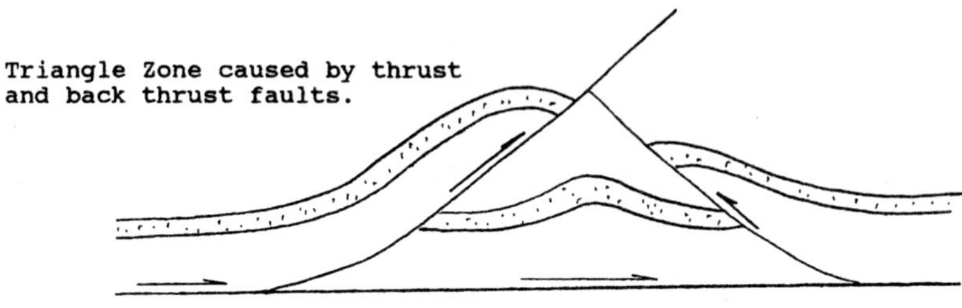

Figure 2.12
Examples of anticlines associated with thrust fault ramps: (a) ramp anticline; (b) imbricate faults (stacked thrust faults—like shingles) and sandwiched duplexes; and (c) triangle zone associated with thrust and back thrust.

associated with ramps include imbricate structures, duplexes, and triangle zones (Fig. 2–12).

2.2.4 Laramide structures

An altogether different geometry exists for thrust faults associated with Laramide deformation in the western United States. Laramide type deformation can be thought of as foreland thrusting, that is, thrusting that occurs in granite cored mountain ranges to the east of the western overthrust belt (Fig. 2–13).

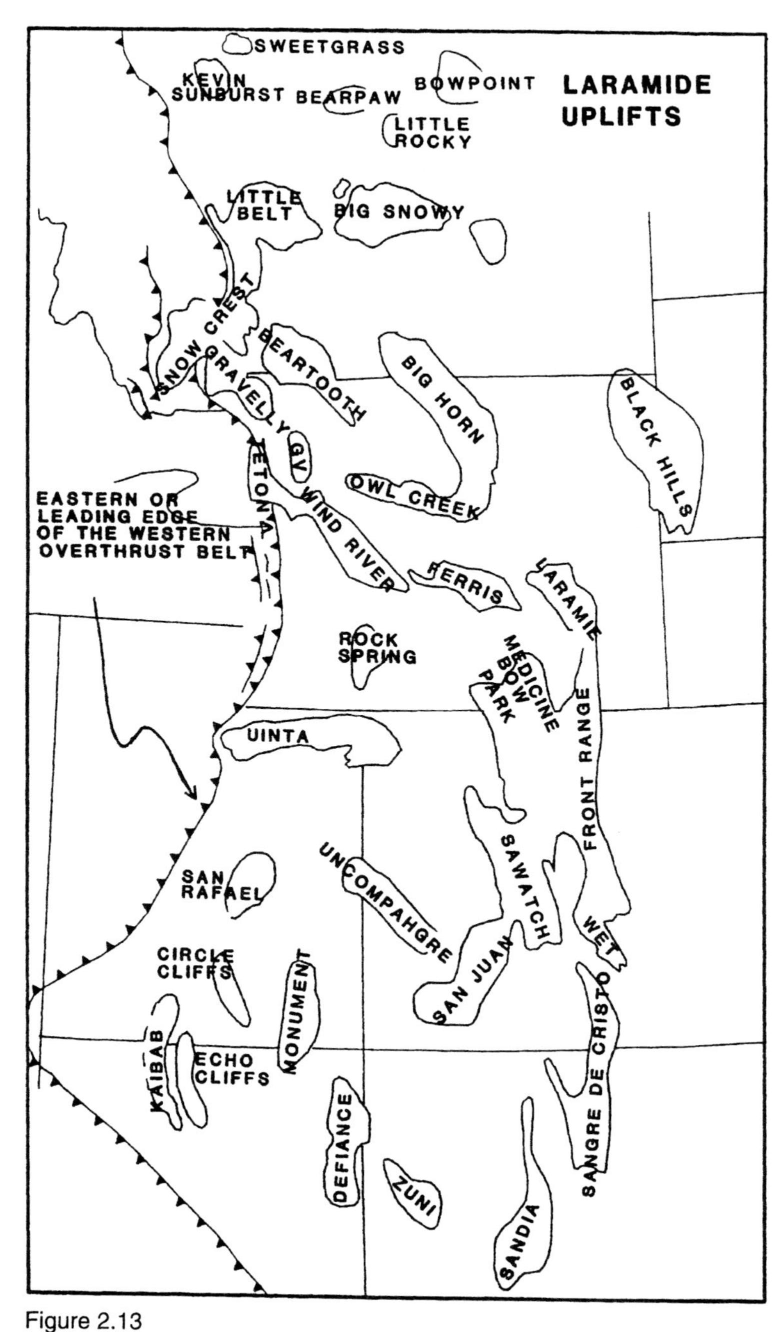

Figure 2.13
Map view of the Western U.S., showing the leading edge of the western overthrust belt and foreland, or Laramide thrust-faulted mountain ranges to the east. Modified from Hintze (1982).

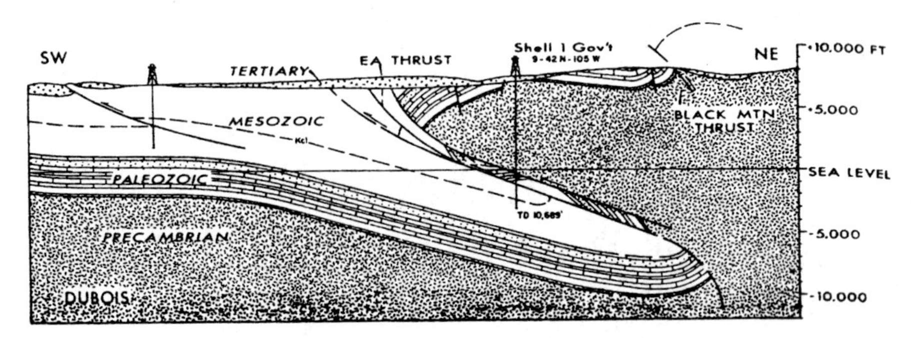

Figure 2.14
NE–SW cross section through a typical Laramide structure near Dubois, Wyoming. Note overturned Paleozoic section near TD of the Shell 1 government well. From Berg (1962), reprinted by permission of AAPG.

Figure 2–14 shows a cross section through a typical Laramide structure near Dubois, Wyoming. The origin of the structures is controversial, but the geometry is such that Precambrian granite is thrust upward (upthrust) and laterally over highly deformed, commonly overturned sedimentary rocks. Where they can be observed, the faults normally dip at about 30 degrees. The trap normally occurs in the sedimentary rocks in the underthrust side of the fault. The trap is commonly an overturned syncline (now antiform) as shown in Figure 2–14. The extraordinary thing about these structures is that to drill such a structure, the well must drill through several thousand feet of overhanging Precambrian granite to reach the sedimentary rocks under the overhang. To date, about eighteen such wells have been drilled with varying results. Though the geology is clearly very complicated, many geologists today feel that these structures have perhaps the best exploration potential in North America for possible future discovery of elephants.

2.2.5 Strike-slip faults

Strike-slip or shear faults are the most important fault types in California. However, in California true fault traps are relatively rare. Most hydrocarbons in California are trapped in anticlines associated with compression of basins or sub-basins located between major fault sets. Blocks of rock located between parallel sets of strike-slip faults are commonly rotated, compressed (transpressed) and warped during shearing. Most California oilfields have complex stratigraphic and structural histories associated with the development of such basins, and it is difficult to make generalizations about them. No overall generalizations such as "roll over into a growth fault" or "leading-edge recumbent anticline" can be made for strike-slip faulted terrains.

Basins associated with major shear zones can be important for accumulation of hydrocarbon-bearing sediments. Many California basins contain very large volumes of very rich source rocks. Parts of the Los Angeles Basin have already produced more than two million barrels per cubic mile of rock in the basin.

True fault traps do occur in such basins. But more commonly, faults are responsible for breaking up of the basin into sub-basins, for rotation, compression, and warping of the sub-basins, for providing hydrocarbon migration pathways from source rocks to traps, and for leaking of hydrocarbons to the surface.

2.2.6 Other fault types

Transform faults associated with spreading centers are not considered important for the development of traps. However, they can play an important role in the development of evaporite basins associated with continental rifting. Evaporites can be extremely important as source rocks and hydrocarbon seals.

2.3 Mapping Faults

Important fault types have been classified here as either normal, reverse, or strike-slip. While there are other fault types, they are of lesser importance in petroleum geology. All faults are curved both in plan view and in cross-section view. All dip-slip faults have some component of strike slip, and all strike-slip faults have some component of dip slip. All faults are three-dimensional features that result from the release of stress in a nonhomogeneous earth. As such, their movement components change both locally and regionally. Some geologists do not like to see a map in which the faults have been constructed using a straight edge because people make straight lines; nature does not.

In mapping on a field scale, it is important that the development geologist attempt to construct structure contour maps on the fault planes themselves. In fields with sparse data or very small faults it is not always possible to do so. But for complex fields with multiple production horizons, fault-plane maps should be constructed because:

1. Dipping faults cut different horizons at different map localities. For a structural interpretation to be consistent at all horizons, fault-cut maps must be prepared.
2. The process of preparing a fault-cut map helps the development geologist understand the field three-dimensionally, and it helps eliminate obvious fault errors.
3. Fault-plane maps define precisely where the faults are (and are not) on a structure contour map of a reservoir. Explorationists commonly don't need to know the exact location of a fault as long as the anticline is there. But the development geologist needs to know exactly where the faults are (and are not) if development wells are to be placed properly in a reservoir.
4. Computers are not capable of properly contouring fields if the locations of the faults are not known. And no computer compares to the human mind in the analysis of the complex and often imprecise data that exists in the interpretation of complex fields.
5. Fault-plane maps are critical for projection of shallow data to deep, untested horizons.

Faults can be contoured by posting fault-cut information including elevations and throws (repeated or missing section) at the well locations on a base map. Faults are correlated usually by throw plus other information, including seismic data. Fault-contour maps are then constructed from fault-cut elevation data. For most normal faults, a 1000-foot contour interval is used, and the 1000-foot contour lines are 700 feet apart because most normal faults dip at about 55 degrees.

By overlaying a contoured fault cut map on a structure contour map of a reservoir, the intersection of common elevations on the two maps precisely defines the location of the fault on the reservoir map. For more details on proper mapping procedures, see Tearpock and Bischke (1991).

2.4 Stratigraphic Traps

In Table 2–1 we have examined different trap types, including stratigraphic traps. It is difficult to make generalities about stratigraphic traps except that, once again, they must close three-dimensionally. The most common stratigraphic trap encountered by the development geologist is a discontinuous reservoir rock such as a channel sandstone, a barrier island sandstone, or carbonate reef or bank. Details of stratigraphic traps, including overall geometries and internal configurations of reservoir rocks are discussed in Chapters 12 and 13 on Clastic and Carbonate Depositional Systems.

2.5 Combination Traps

Finally, many traps are either combination traps or defy classification. Prudhoe Bay (see Chapter 12), the largest oilfield in North America, is a combination faulted anticline-angular unconformity trap. Panhandle-Hugoton (Pippin, 1970), the largest gas field in North America, is a combination hydrodynamic-stratigraphic pinch-out anticline.

2.6 Salt Domes

In addition to being badly faulted, many salt domes have stratigraphic irregularities on their flanks. Salt domes normally go through three stages of development (Fig. 2–15), and the identification of each stage can be important in locating potential stratigraphic traps. During stage 1, the pillow stage, very subtle stratigraphic pinch outs can be noted at substantial distances from the dome. During stage 2, the true piercement stage, the salt physically intrudes overlying rock, and stratigraphic anomalies occur only locally and at the surface. During stage 3, the post-diapiric stage, long periods of down-building sedimentation occur on the flanks of the dome, and local stratigraphic anomalies can occur throughout large vertical sequences. Figure 2–16 shows most of the different types of traps that can occur in the vicinity of salt domes.

2.7 Other Trap Types

Oil and gas are sometimes produced from reservoirs where the actual trap configuration has not or cannot be defined. In some cases these are fractured shales that occur deep within the synclinal cores of basins. In other cases they seem to be reservoir rocks with no definable outline as to the limits of the reservoir. Many are thought to occur in areas where oil and gas are being generated actively. That is, they are their own source rock, and the oil and gas have simply never been concentrated in a definable trap. Parts of the Austin Chalk of Texas, Codell Formation of Colorado, Monterrey Formation of California, and the Wasatach Formation of the Altamont-Bluebell Field in Utah contain such reservoirs.

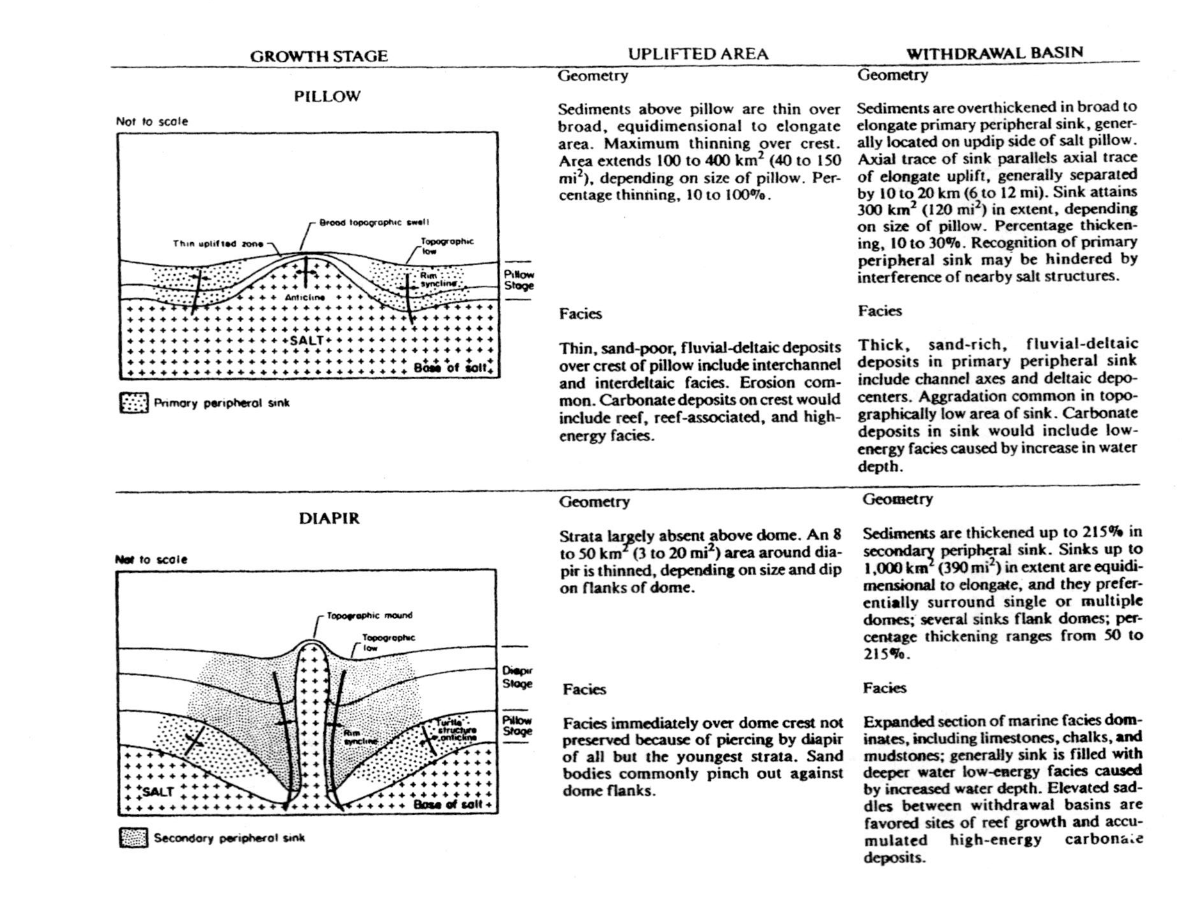

GROWTH STAGE	UPLIFTED AREA	WITHDRAWAL BASIN
PILLOW	Geometry Sediments above pillow are thin over broad, equidimensional to elongate area. Maximum thinning over crest. Area extends 100 to 400 km^2 (40 to 150 mi^2), depending on size of pillow. Percentage thinning, 10 to 100%. Facies Thin, sand-poor, fluvial-deltaic deposits over crest of pillow include interchannel and interdeltaic facies. Erosion common. Carbonate deposits on crest would include reef, reef-associated, and high-energy facies.	Geometry Sediments are overthickened in broad to elongate primary peripheral sink, generally located on updip side of salt pillow. Axial trace of sink parallels axial trace of elongate uplift, generally separated by 10 to 20 km (6 to 12 mi). Sink attains 300 km^2 (120 mi^2) in extent, depending on size of pillow. Percentage thickening, 10 to 30%. Recognition of primary peripheral sink may be hindered by interference of nearby salt structures. Facies Thick, sand-rich, fluvial-deltaic deposits in primary peripheral sink include channel axes and deltaic depocenters. Aggradation common in topographically low area of sink. Carbonate deposits in sink would include low-energy facies caused by increase in water depth.
DIAPIR	Geometry Strata largely absent above dome. An 8 to 50 km^2 (3 to 20 mi^2) area around diapir is thinned, depending on size and dip on flanks of dome. Facies Facies immediately over dome crest not preserved because of piercing by diapir of all but the youngest strata. Sand bodies commonly pinch out against dome flanks.	Geometry Sediments are thickened up to 215% in secondary peripheral sink. Sinks up to 1,000 km^2 (390 mi^2) in extent are equidimensional to elongate, and they preferentially surround single or multiple domes; several sinks flank domes; percentage thickening ranges from 50 to 215%. Facies Expanded section of marine facies dominates, including limestones, chalks, and mudstones; generally sink is filled with deeper water low-energy facies caused by increased water depth. Elevated saddles between withdrawal basins are favored sites of reef growth and accumulated high-energy carbonate deposits.

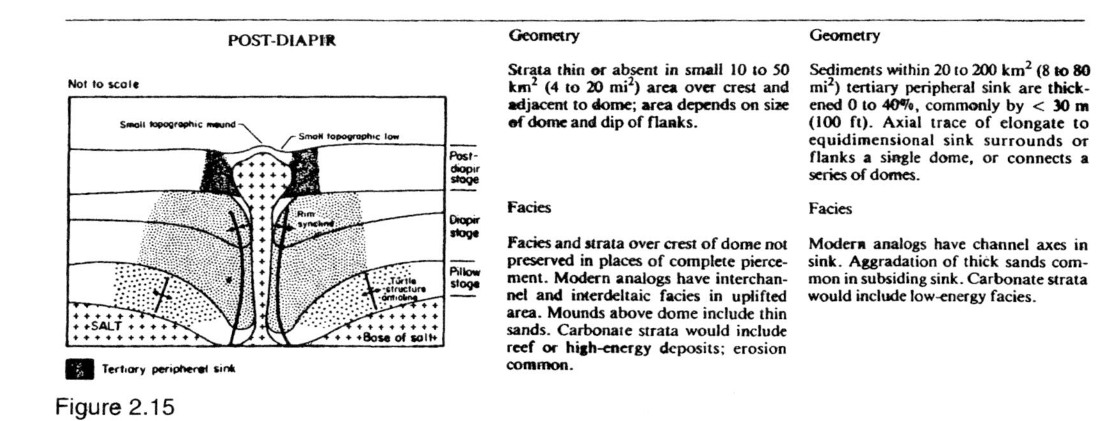

Geometry	Geometry
Strata thin or absent in small 10 to 50 km^2 (4 to 20 mi^2) area over crest and adjacent to dome; area depends on size of dome and dip of flanks.	Sediments within 20 to 200 km^2 (8 to 80 mi^2) tertiary peripheral sink are thickened 0 to 40%, commonly by < 30 m (100 ft). Axial trace of elongate to equidimensional sink surrounds or flanks a single dome, or connects a series of domes.
Facies	**Facies**
Facies and strata over crest of dome not preserved in places of complete piercement. Modern analogs have interchannel and interdeltaic facies in uplifted area. Mounds above dome include thin sands. Carbonate strata would include reef or high-energy deposits; erosion common.	Modern analogs have channel axes in sink. Aggradation of thick sands common in subsiding sink. Carbonate strata would include low-energy facies.

Figure 2.15
Three stages of salt dome development. From Seni & Jackson (1983a), reprinted by permission of AAPG.

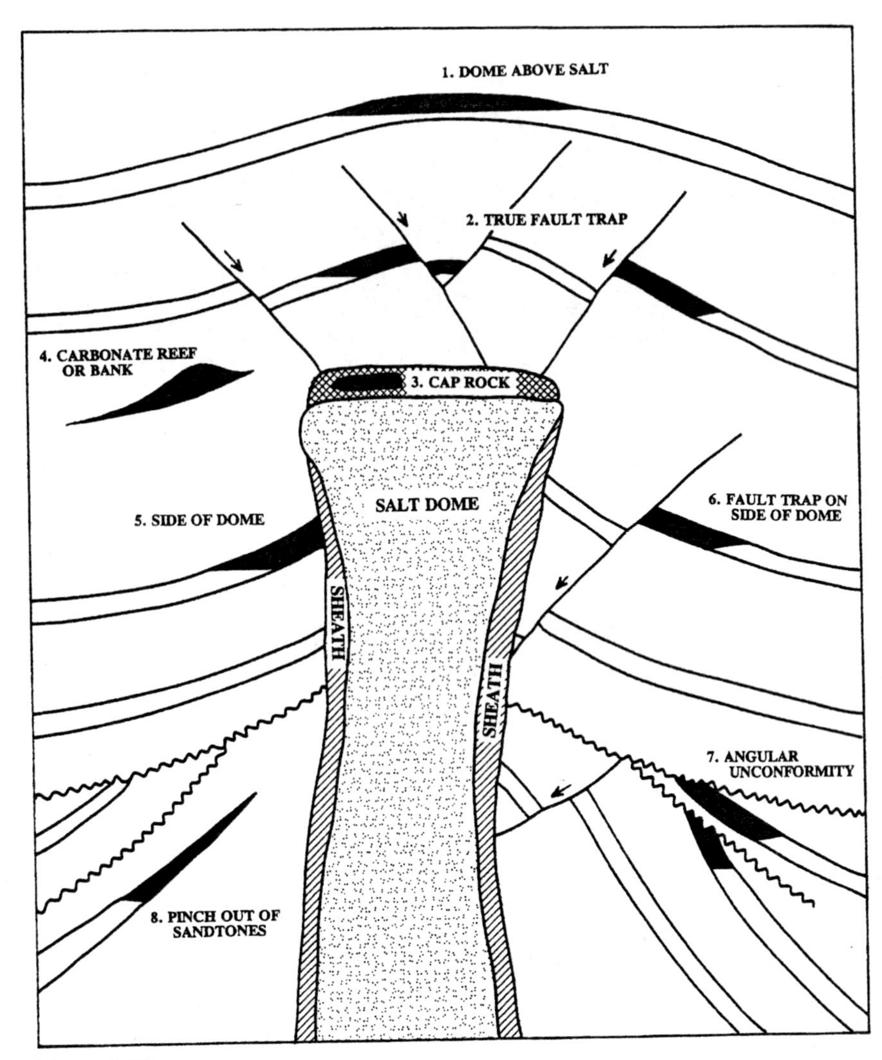

Figure 2.16
Idealized cross section through a salt dome, showing eight different traps associated with the dome.

CHAPTER 3

Reserve Estimates

Young geologists are sometimes under the misconception that, if the geology is good, the well will get drilled. In fact, wells don't get drilled just because the geology is good; they get drilled because the geology is good *and* because there is potential for economic gain. Potential income from any petroleum venture is based on the oil or gas that is expected to be produced *in the future*. Perhaps the single most important charge of the development geologist is to estimate the oil and gas reserves that may be discovered by any particular venture, and to keep track of the reserves in all past ventures.

3.1 Four Basic Methods

Reserve estimates are normally made by four different methods:

1. Educated estimate (guess)
2. Comparison of nearby, similar production histories
3. Volumetric calculations
4. Reservoir simulation—material balance calculations

3.1.1 Educated estimate

In the good old days many wells were drilled simply because people thought that there might be oil on the property or lease. Economic analyses were not run. People just drilled, and if they found something, great. If not, they moved on. Many of the largest discoveries in the nation were drilled on this basis. *Spindletop,* by Clark & Halbouty (1952), is a fantastic story of wildcatters drilling wells where they had no idea what was going to come out of the ground. *The Last Boom,* also by Clark & Halbouty (1972), a history of the East Texas field, is an unbelievable story about a nongeologist who discovered the largest oilfield in North America (until the discovery of Prudhoe Bay) because he thought that there was "an ocean of oil" under Rusk County, Texas. In discovering this field, Dad Joiner confirmed one of the most important concepts in petroleum geology, the stratigraphic trap. Even today, some wells are drilled without economic analyses mainly because of contract obligations, or an owner might keep a rig running at a loss rather than stack, or store, the rig.

However, most of these wells are drilled by individuals or companies already committed to drilling a number of wells. It is unlikely that, as a company geologist, you will get a chance to drill a well because you feel that "there is an ocean of oil under some property" unless you want to pay for it yourself.

3.1.2 Comparison of nearby production histories

In reservoirs where porosity and permeability distributions are unpredictable and volumetric calculations have little meaning, such as in the fractured Austin Chalk of Texas or tight sandstones of the Codell Formation of the Denver-Julesburg basin, one way to estimate the potential production for a proposed well is to analyze production histories of nearby wells. Generally, a wildcat well will not produce more than the average production from the five or ten closest wells. Even though it is possible that an exploratory well may produce more than any nearby wells, your credibility may suffer if you assign potential reserves larger than the average of nearby wells.

3.1.3 Volumetrics

Volumetric estimates of reserves are the most accurate and widely used method of estimating potential reserves in exploration and development situations. Volumetrics are the domain of the geologist because volumetric calculations are based primarily on the geologist's structure and isopach maps. Volumetric estimates form the basis for most reserve calculations in oil companies today, and the majority of this chapter will be devoted to volumetrics.

3.1.4 Reservoir simulation

Reservoir modeling is primarily the domain of the reservoir engineer with input from the geologist. A number of different model types exist. Probably the most common reservoir simulation model used by the industry today is a *finite difference model.* The reservoir is modeled mathematically in terms of overall shape and internal configuration, including porosity, permeability, fluid saturations, pressure, and internal permeability barrier configurations. Reservoir conditions are simulated using different flow rates, and flow configurations, as well as different injection types and configurations. Internal reservoir conditions are calculated using material balance equations, and different development schemes can be compared for optimum development plans. The principal problem with this procedure is that truly accurate models cannot usually be developed until after development drilling has occurred and the reservoirs have been on-line for a period of time.

After a well or group of wells has produced from a particular reservoir for an unspecified length of time, a decline rate can often be established. When a reservoir has been on-line for long enough to establish a decline rate (commonly six months to a year), reasonably good estimates of the reserves can be made through the use of decline curves. These techniques are the domain of the petroleum engineer, and the reserves generated by these techniques are commonly used for comparison against volumetric calculations. Neither is considered more accurate than the other. When discrepancies are noted, it is time for the geologist and the engineer to get together to reconcile the differences. Chapter 11 is devoted entirely to decline curves, one of the simplest and most important material balance-reservoir simulation techniques.

3.2 Volumetrics

The remainder of this chapter will be devoted to volumetrics, the most important charge of the development geologist. In one way or another, the development geologist must establish rock volumes that are assumed to contain oil and/or gas. This calculation may be no more elaborate than an area, as determined by a seismic bright spot, times some reservoir thickness. Or it may be as elaborate as a net gas or net oil isopach, as determined by structure contour maps modified by fluid contacts and net reservoir thickness isopachs.

Most rock volumes are established through the use of net gas and net oil isopach maps. These maps are made from structure contour maps with defined gas-oil and/or oil-water contacts. Laboratory mapping exercises should be used to establish these procedures.

Once the rock volume has been established, then the in-place oil and/or gas can be calculated by:

1. Determination of pore volume which is equal to the rock volume times an average porosity. Average porosity numbers are commonly calculated by well-log analysts (petrophysicists), geologists, or petroleum engineers (see Chapter 4).
2. Determination of the in-place gas and/or oil volume by subtracting out the water saturation, connate, or free water that exists in all reservoir rocks. Water saturation numbers are normally calculated by well-log analysts, geologists, or petroleum engineers (see Chapter 4).
3. Correcting to sales line temperature and pressure through the use of formation volume factors. Formation volume factors are normally determined by reservoir engineers (see Chapter 8).

Notice that in-place reserves are not the same thing as recoverable reserves. This may seem like a trivial point, but it is not uncommon for explorationists to attempt to sell in-place reserves. Recoverable reserves are the percentage of in-place reserves that are expected to be recovered as estimated by a *recovery efficiency*. Recovery efficiencies are dependent on many factors and will be discussed in detail in Chapter 9. However, the world average recovery efficiency for primary oil recovery is approximately 35 percent. If you are buying oil reserves, let there be no misunderstanding as to whether you are buying in-place or recoverable reserves.

Note also that all reserves, whether in-place or recoverable, are expressed in surface, or pipeline units. It does no good to talk about a cubic foot of gas at reservoir conditions because one cubic foot of gas in reservoir A at 5,000 psi will contain on the order of ten times as much gas as one cubic foot of gas in reservoir B at 500 psi. Each must be converted to a common sales pressure base before their volumes have any meaning. The same is true for oil which almost always shrinks in travelling from reservoir conditions to the surface.

Table 3–1 summarizes the most commonly used formulas for volumetric reserve calculations. Many of the formulas have grown from field procedures and field personnel. As a result, some peculiar units and conversion factors will be noted.

Table 3–1 Summary of formulas for volumetric reserve calculations.

$$\text{Recoverable gas} = \text{rock volume} \cdot \frac{43{,}560 \cdot \phi \cdot S_{hc}}{FVF_{gas}} \cdot RE$$

$$\text{Recoverable oil} = \text{rock volume} \cdot \underbrace{\underbrace{\frac{7{,}758 \cdot \phi \cdot S_{hc}}{FVF_{oil}}}_{\text{in-place factors}} \cdot RE}_{\text{recovery factors}}$$

rock volume is from an isopach map, or some area times a thickness

43,560	is a conversion factor from acre feet to cubic feet
7,758	is a conversion factor from acre feet to barrels
ϕ	is porosity as a decimal from well logs or cores
S_{hc}	is hydrocarbon saturation or $(1 - S_w)$ where S_w is water saturation from well logs and the Archie Equation
FVF	is formation volume factor expressed in reservoir units/surface units
FVF_{gas}	Because gas always expands in travelling from reservoir conditions to surface conditions, FVF_{gas} is always less than 1. For dry gas FVF_{gas} can be estimated by using 15 psi/reservoir pressure. FVF_{gas} is primarily a function of reservoir pressure, pressure base, and gas composition.
FVF_{oil}	Because oil usually shrinks in travelling from reservoir conditions to surface conditions, FVF_{oil} is always equal to or greater than 1. It is commonly in the range of 1.1 to 1.5.
RE	stands for recovery efficiency and is an educated estimate. It depends on porosity, permeability, composition of the oil or gas, well spacing, drive mechanism, plus secondary and tertiary recovery techniques employed. The world average RE for gas is 70 percent. The world average RE for primary plus secondary oil recovery is 35 percent.

Useful equivalents:

One hectare = 0.01 km^2 or 10,000 m^2 or 2.471 acres

One mile = 5,280 feet

One square mile = 640 acres

One acre = 208.7 feet on a side

160 acres = ¼ mile square

One meter = 3.281 feet

One barrel = 42 U. S. gallons = 35 imperial gallons = 5.615 cubic feet

One cubic meter = 6.29 U. S. barrels = 35.320 cubic feet

3.3 Uncertainty

Regardless of which type of reserve estimation is made, there is always some uncertainty about the accuracy of the estimate. Although geologists must make one final structural and stratigraphic interpretation, they are allowed to hedge their bets by building minimum, most likely, and maximum case scenarios, particularly if fluid levels have not been identified or the reserves are based on bright spot technology.

This uncertainty may be approached by a number of different methods. Some companies have their own internal methods of dealing with uncertainty, but most have adopted the terminology set forth in the *Guidelines for Application of the Definitions for Oil and Gas Reserves,* Monograph I, by the Society of Petroleum Engineers (SPE, 1988), including complete definitions and a thorough discussion of the terms. The following is an abbreviated discussion of reserve terms.

The term *reserve* refers to commodities that may be recovered economically under current economic and technologic conditions. Reserves may be classified as either proved or unproved.

Proved reserves can be estimated with reasonable certainty to be recoverable under current economic conditions. In certain cases, reserves may be classified as proved purely on the basis of electric logs, but normally reserves are considered proved only if commercial producibility of the reservoir is supported by actual production or formation tests. Proved reserves may be developed or undeveloped.

Proved developed reserves are reserves that are expected to be recovered from existing wells. *Proved undeveloped reserves* are to be recovered from new wells in undrilled acreage, deepening wells to different reservoirs, or when additional expenditures will be required to obtain production.

Unproved reserves are based on data similar to that used in estimates of proved reserves, but technical, contractual, economic, or regulatory uncertainties prevent these reserves from being classified as proved. Unproved reserves may be probable or possible.

Probable reserves are less certain than proved reserves and can be estimated with some reasonable degree of certainty; they may include reserves estimated for stepout drilling, reserves based on logging estimates, or reserves available from improved recovery methods not currently in operation.

Possible reserves are not as certain as probable reserves and can only be estimated with a low degree of confidence.

Although these terms are now well-defined, a great deal of latitude still exists between and sometimes within companies in their usage. Three-dimensional seismic and bright spot technology has added a new and poorly defined dimension to reserve estimation. For decision analysis, the following generalizations can be made:

1. *Proved reserves* normally define minimum case economics for a venture. Given the uncertainty that exists in a particular venture, the venture should do no worse than the minimum case estimate. Financial agencies will normally loan money using proved reserves as collateral.
2. *Proved plus probable reserves* normally define the *most likely* or *best guess* occurrence in a probabilistic sense. Most companies make internal decisions based on the most likely probabilistic case, which is proved plus probable reserves.
3. *Proved plus probable plus possible reserves* usually define the maximum economic case. Given the uncertainty that exists for any given venture, this is the best that could reasonably happen. Most companies try to sell reserves based on the maximum case scenario.

The terms *reserve* and *resource* are commonly used interchangeably and improperly in the industry. A *resource* is any commodity that is now or may become economic in the future. It includes commodity volumes that may be undiscovered as yet. A *reserve* is a natural resource that may be recovered economically under today's economic and technologic conditions. Thus, the term *potential reserve* is self-contradictory, but is used quite commonly in the business.

An example of the problem exists in federal waters off the east coast of the U.S., where a significant gas discovery was made several years ago. Exploratory drilling delineated and "proved" that significant quantities of gas were present, but the volume of gas discovered was not enough to justify the cost of a pipeline to the discovery. Thus, a gas volume has been proved, but it is not a reserve because it is not economic. It is a resource that may, in the future, become a reserve. Is it a potential reserve?

Commonly, decisions in a major oil company are based on diagrams similar to the one shown in Figure 3–1. This is an example of an offshore case where a company is attempting to determine whether it should spend fifty million dollars to proceed with the development of a field by

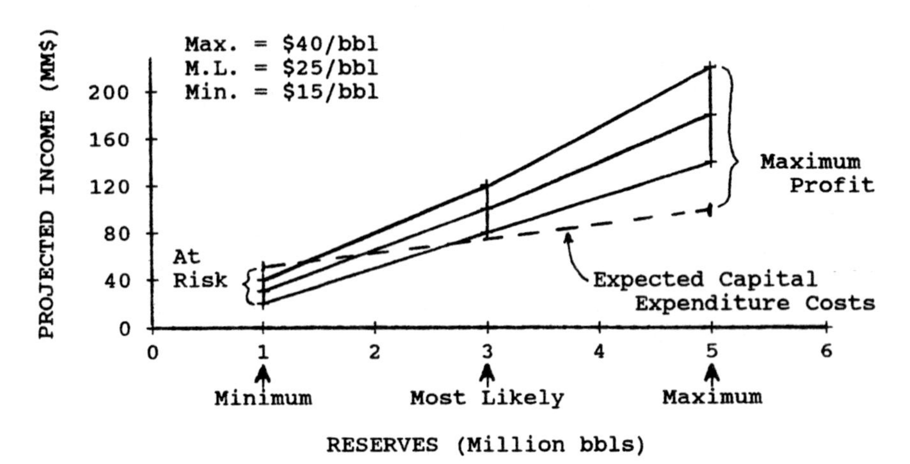

Figure 3.1
Example of comparison between capital cost and potential income for uncertain reserves and uncertain price projections. Diagram illustrates potential profit relative to risk for decision analysis under uncertain conditions.

the building and setting of an offshore platform. Based on already drilled exploratory wells, the geologists has determined that the field contains recoverable reserves estimated at:

Maximum case = 5,000,000 barrels (proved + probable + possible)
Most likely case = 3,000,000 barrels (proved plus probable)
Minimum case = 1,000,000 barrels (proved)

Economists may also hedge their bets by estimating that when the oil is produced the oil will be worth:

Maximum case = $40/barrel
Most likely = $25/barrel
Minimum = $15/barrel

Thus, an expected income envelope can be developed from the diagram based on expected reserves. If the expected cost for the platform, pipeline, development wells, scrubbers, dehydrators, separators, collection facilities, pig launchers, and all the equipment that goes on offshore platforms is estimated at $50,000,000 (increasing slightly as reserves increase), then one can look at the diagram and develop some idea as to the upside and downside potential to the venture.

Notice that the minimum case reserves are not really reserves because they are not economically extractable. In this case, if the company proceeds with the venture and the minimum case reserves prove to be correct, then the company will lose an estimated $10 to $25 million. If the most likely reserves prove to be correct then the company will make a profit, and so on.

In this case, even though the upside potential greatly exceeds the downside potential, the company would probably not proceed with development at this point. If possible, the company would probably drill another exploratory well which would either raise the minimum case reserves to the break-even point or lower the most likely and maximum case reserves to below the break-even point. Sometimes this can be done; sometimes it cannot. What the company does not want to do is spend $500,000 for another exploratory well only to have the estimates remain unchanged.

The decision analysis game is a fascinating and complicated one in which the geologist's reserve estimates are commonly the most important factor in the decision. It is also important to note that these reserve estimates are the development geologist's numbers, not the exploration department's. In most companies, the production department not the exploration department decides whether to set a platform.

3.4 Isopach Maps

A fair amount of confusion exists regarding the terms isolith, isopach, and isochore. *Isopach* maps represent graphic representations of the vertical thickness of some parameter, such as the vertical thickness of the reservoir rock, the vertical thickness of reservoir rock containing gas, or the vertical thickness of rock saturated with oil. *Isochore* maps are usually used to show convergence between two or more stratigraphic horizons. If the stratigraphic horizons happen to be the top and bottom of a reservoir, the two terms may mean the same thing.

Isolith maps represent the true stratigraphic thickness of a lithologic horizon. The distinction between isolith and isopach is not usually very important until high dips are encountered, when the distinction becomes very important. In the estimation of reserves it is far easier to use isopach maps than isolith maps, because an isopach map is projected directly to the flat map surface to determine area, while an isolith map must be rotated to determine its surface area. Figure 3–2 illustrates the difference.

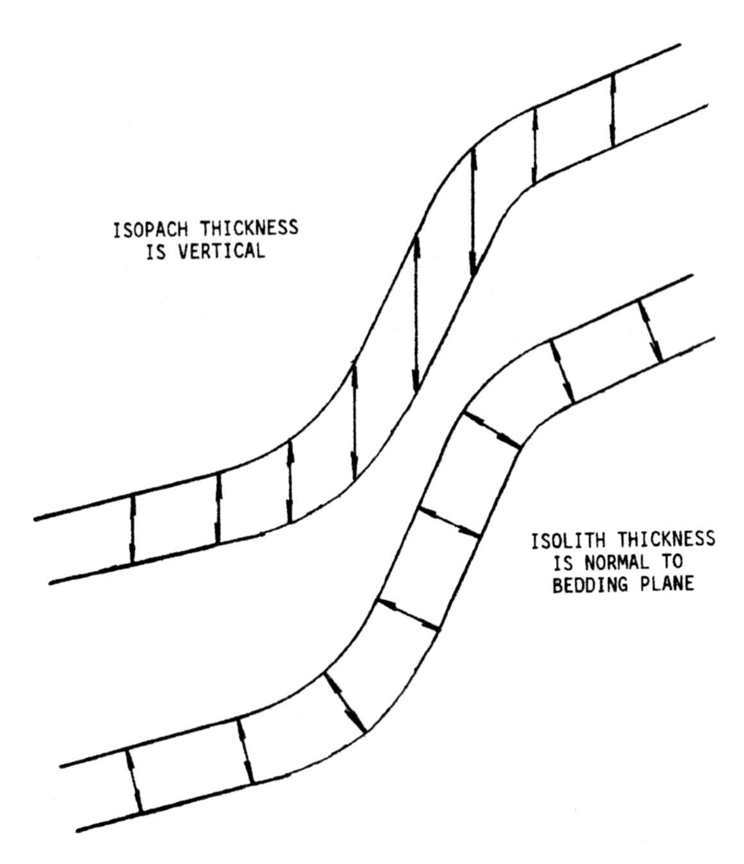

Figure 3.2

Cross section showing difference between isopach and isolith thickness. In this diagram isolith thickness is constant; isopach thickness is not.

Isopach maps are used for volumetric reserve estimates because they overlay structure contour maps and represent vertical thicknesses of rock volume saturated with either oil or gas. The areal extent of the various thicknesses are used to define rock volumes that are saturated with hydrocarbons. Many young, uninitiated geologists will simply take the net gas or net oil data from the pay counts and draw an isopach around the data. This is not appropriate for any reservoir that has well-defined fluid contacts.

In designing an isopach, several factors must be considered. If there is a great deal of data and the reservoir is irregular, the isopach should have a relatively small contour interval and the areas should be planimetered or calculated by digitizing data on the computer. If there is very little data, and/or the reservoir is considered to be regular, a large contour interval can be used and some shortcuts can be taken (as described later in this chapter).

3.4.1 Gross sand thickness isopachs

Gross sand thickness refers to the total interval of rock saturated with oil or gas irrespective of tight zones, zones of low porosity or low permeability. Gross sand thickness isopachs are quite easy to make, especially for gas. The zero line for the isopach corresponds to the downdip limits of gas, the gas-oil contact if oil is present, or the gas-water contact if no oil is present. Gross isopach thickness lines should increase updip in a fashion that corresponds exactly to the appropriate structure contour elevations. For example, if the gas-water contact is at –7,000 feet subsea, then the following isopach lines should exactly overlay the following structure contour lines:

Structure Contour Line	Gross Isopach Line
–7,000	0′
–6,980	20′
–6,960	40′

and so forth until the sand becomes full or the top of the structure is reached. Within this maximum thickness area either a maximum thickness or an average thickness should be recorded on the isopach. This maximum thickness will typically not correspond to one of the isopach contour line thicknesses.

For gross oil isopachs, the sand thicknesses should increase updip from the oil-water contact exactly as described for the gas reservoir. However, if gas is present, then a decreasing thickness wedge will occur updip of the gas-oil contact.

3.4.2 Net pay thickness isopachs

Net thickness refers to the gross reservoir thickness, with tight zones (low-porosity zones) thrown out. For example, a sandstone that is 100 gross feet thick might have 20 feet of rock that has low porosity, considered to be nonreservoir rock. Such a sandstone has 80 net feet of reservoir rock and has a net/gross ratio of 80 percent.

There are several ways to deal with such a reservoir. If the net/gross ratio across the reservoir is uniformly 80 percent, then it may be valid to make gross gas and gross oil isopachs and simply multiply the resulting rock volumes by 0.8 to arrive at the appropriate net rock volumes.

Alternatively, a net gas isopach and net oil isopach can be prepared by simply adjusting the isopach thickness by the reciprocal of the net/gross ratio. In the above example the 20-net foot thickness line should coincide with the –6,975 elevation on the structure contour map rather than the –6,980 line.

More commonly, where many wells are present, the isopach thicknesses are drawn parallel to the structure contour lines, but are adjusted to honor the individual well pay counts.

3.4.3 Variable reservoir thickness

Where the vertical thickness of the reservoir changes laterally, such as at the edge of a channel or reef, a net reservoir thickness isopach may be necessary.

The downdip limits for the oil and gas isopach are easily identified by overlaying the net sand thickness isopach on the structure contour map. Again, the zero gas isopach line coincides with the gas-water or gas-oil fluid level, as identified on the structure contour line. But the zero gas isopach line departs from the fluid level line as it intersects the zero sand thickness line on the reservoir thickness isopach map. Recoverable oil and gas cannot exist where there is no reservoir rock. The normal net gas or net oil isopach lines simply switch from the fluid contact wedge lines to the reservoir thickness wedge lines when the sand becomes full of oil or gas. In other words, the normal net oil or gas isopachs must be modified such that the net thickness of oil or gas does not exceed the net thickness of the reservoir. This will probably become clear only through laboratory examples.

Again, the updip limits for oil are confusing. The zero line for the updip limit of oil is defined by the sand-full-of-gas line, which is defined by the intersection of the sand wedge isopach lines with the gas-oil wedge isopach lines on the gas isopach. The oil isopach thickens downdip, but isopach lines will not parallel the structure contour lines. This will only become clear through an example. The combinations are limitless when the thickness of the reservoir varies.

3.5 Calculations from Isopachs

3.5.1 Trapezoidal rule

An isopach is a graphic representation of rock saturated with oil or gas. The most commonly used method to calculate a rock volume from an isopach is through the *Trapezoidal Rule,* although other methods such as Simpson's Rule and shortcut methods can be employed. The Trapezoidal Rule is written:

$$BV = (h/2)\,[A_0 + 2A_1 + 2A_2 + \ldots + 2A_{n-1} + A_n] + h_n A_n/2$$

where:

BV = bulk volume, typically in acre feet

h = contour interval

A_0 = area enclosed by zero contour line

A_1 = area enclosed by first contour line above zero

A_{n-1} = area enclosed by first contour line below top contour line

A_n = area enclosed by top contour line

h_n = vertical distance from top contour to top of reservoir

Although the Trapezoidal Rule looks fairly complicated, it is doing nothing more than taking the average area between each two levels and multiplying that average area by the contour interval thickness to arrive at a volume (see Fig. 3–3).

3.5.2 Shortcuts

Isopach maps tend to fall into three general categories. Two of those types offer shortcut methods that should be used routinely as a check against detailed isopachs.

The three types are:

1. **Top of a sphere** (see Fig. 3–3): The volume of the top of a sphere is remarkably close to the area of the base times one half the maximum thickness (or height). This is nothing other than the Trapezoidal Rule applied to one single reservoir thickness (h_n). The same rule applies to the top of an ellipsoid, the top of a

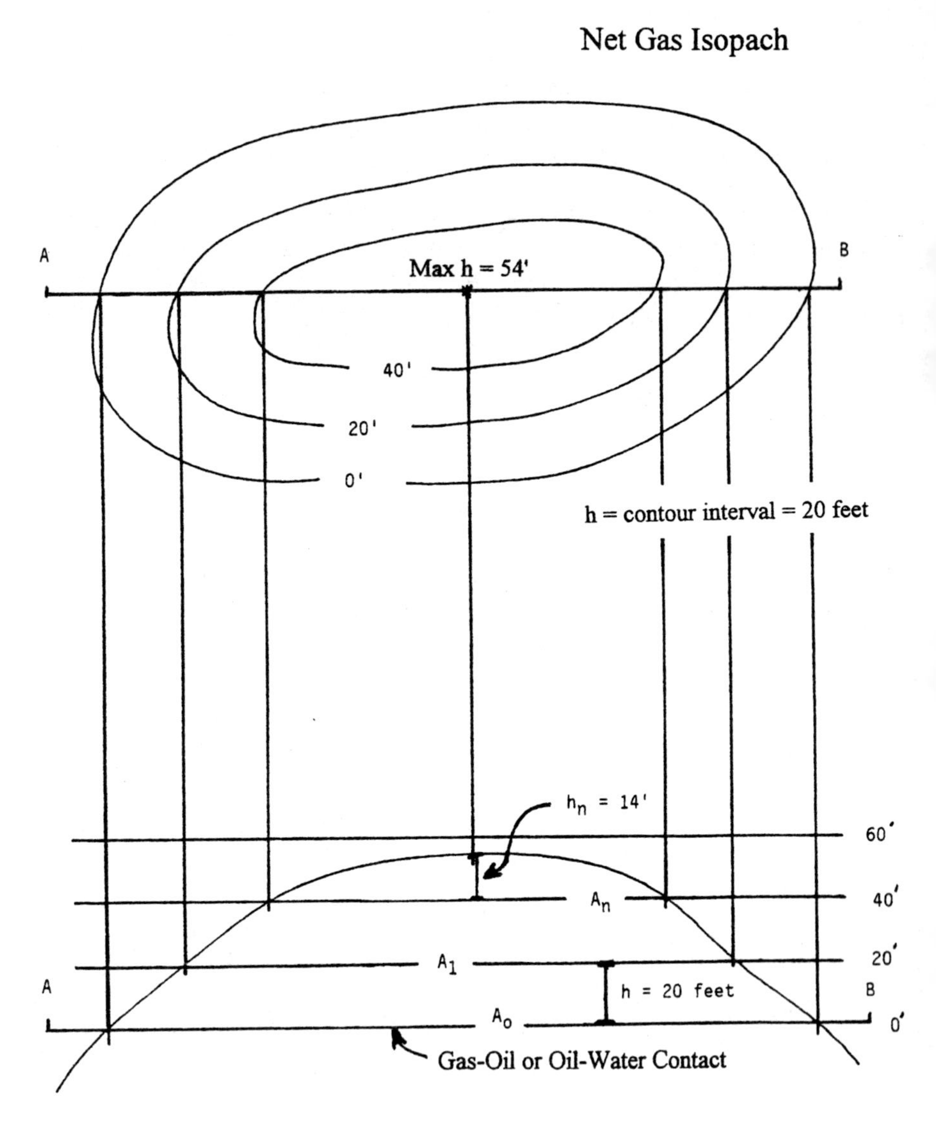

Figure 3.3
Example net gas isopach over the top of a dome (sphere) illustrating the trapezoidal rule. A remarkably accurate shortcut is to multiply the base area (A_0) times one half of the maximum thickness (1/2 of Max h of 54 feet = 27 feet).

faulted sphere, or the top of a faulted ellipsoid. For a perfect sphere this rule is accurate to within 6 percent.

2. **Constant thickness reservoir with wedges on one or more of the margins of the reservoir** (see Fig. 3–4): The wedges are typically the gas-oil wedge or the oil-water wedge but can be other types. The shortcut method here is to average the base area with the maximum thickness area and multiply this average by the maximum thickness. An alternative is to measure the area of the isopach line that corresponds to half of the maximum thickness and multiply this area by the maximum reservoir thickness area.
3. **Irregular shape:** Unfortunately, there are no shortcuts for unusually shaped reservoirs. Fortunately, these are rather rare.

3.6 Measuring Areas

Three methods are generally used for measuring the areas themselves: square counter grid, planimeter, and digitizing on the computer.

3.6.1 Square counter grid

The easiest and least accurate method is to simply overlay a grid of known areas (square counter) and count the area within each of the contour lines. Be careful of scales.

3.6.2 Planimeter

Planimeters are devices that mechanically measure areas inside circumscribed lines. Planimeters should always be calibrated before use, and they should be checked carefully to make certain that all wheels fit tight in their mountings. If the wheels are at all loose, results will be unreliable. Also, it is important to keep the planimeter arms as close as possible to 90 degrees. Do not expect accurate results if the arms deviate by more than 45 degrees from 90 degrees (don't let the arms stretch out to near 180 degrees or back to near 0 degrees. Use these steps to calibrate:

1. Set the planimeter to zero.
2. Move the planimeter clockwise around a known area (such as 40 acres), staying on the lines and returning to the exact starting point.
3. Record the number on the planimeter (for example, 1.4).

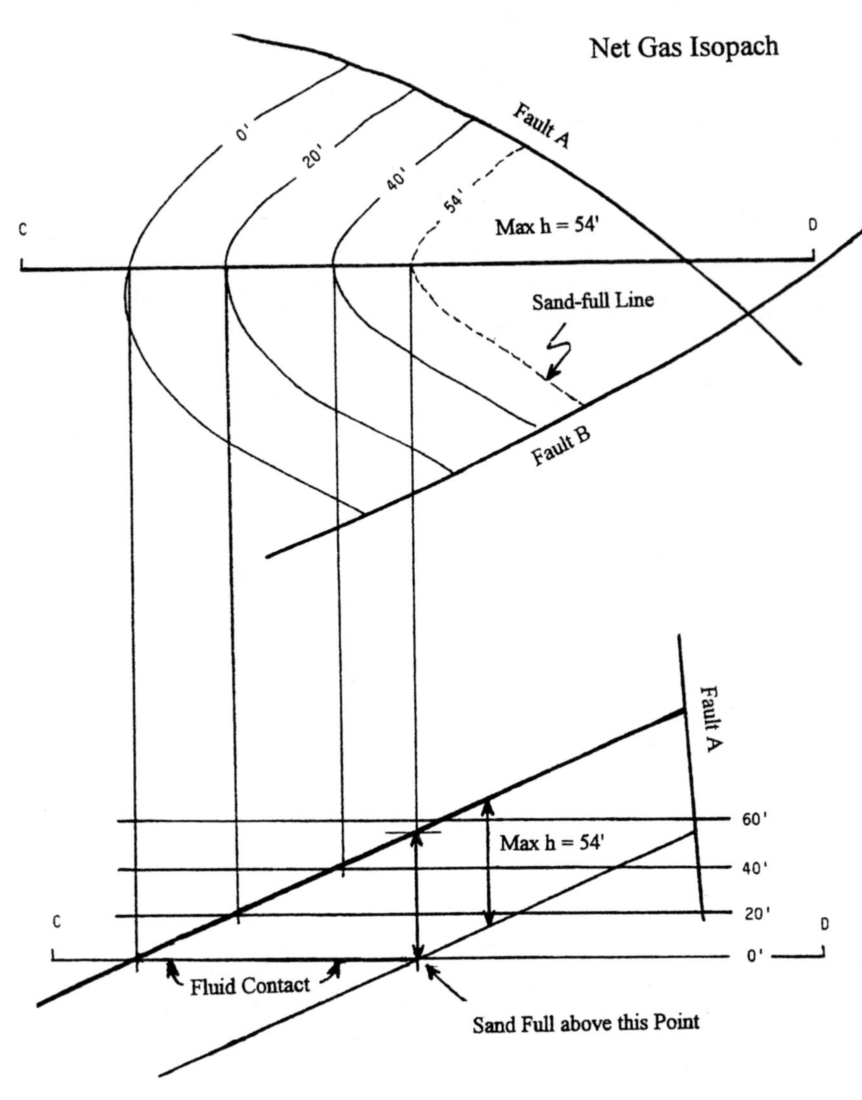

Cross Section through the Above Isopach

Figure 3.4
Example gas isopach where the sand fills to the base of the sand. Above the sand-full line (dotted), the reservoir is full of gas to the base of the sand. Below the sand-full line, a fluid wedge is present. A well drilled in the wedge area will encounter gas plus either oil or water.

A shortcut method for this isopach is to average the A_0 area with the A_{54} area and multiply times the full 54-foot thickness. Another alternative is to multiply the area corresponding to the A_{27} line (half-way, not shown) times the Max h of 54 feet.

4. You now have a ratio (1.4 units equals 40 acres) such that unknown areas can be measured and compared against that ratio.
5. For example, if the unknown area measures 2.1 units, then the area can be calculated using:

$$2.1 \text{ units} \cdot 40 \text{ acres} / 1.4 \text{ units} = 60 \text{ acres}$$

It is very easy to make mistakes with a planimeter, and planimetered areas should always be measured several times and be checked by square counter grids and/or shortcut methods.

3.6.3 Digitizing on the computer

Numerous computer programs can calculate areas very accurately. Isopach lines are simply digitized into the computer using a cursor, and the computer uses Simpson's Rule or the Trapezoidal Rule to calculate the volumes. Computers can make very unusual and seriously inaccurate projections into maximum thickness areas and areas of sparse data. It is also very easy to make scale and/or unit mistakes. Digitized data should also be checked with square counter grids and/or shortcut methods.

CHAPTER 4

Electric Logs

The most important source of data for the development geologist is electric well logs. This chapter covers only the most fundamental concepts of well logs, those that are essential for volumetric reserve calculations. This discussion is not intended to cover logs in a comprehensive manner, and it does not cover many of the specialty logs that can be important to the development geologist. Be aware that exceptions exist for almost every generalization, and specialists should be consulted for anything unusual.

4.1 Datum and Depths

Before a well is drilled, its exact location is surveyed to make certain that it is being drilled at the correct location. In all except some very old wells the survey includes elevation above sea level, and recorded at the top of all modern logs is the datum above sea level from which the log was measured. The *datum* is usually the *drill floor (DF) elevation* or the *Kelly bushing (KB) elevation*. The Kelly bushing is a device that sits on the rotary table and turns the drill pipe. The Kelly bushing is usually about one foot higher than the drill floor. As the well is being drilled, the driller measures the length of the individual stands of pipe and keeps track of the depth at which the well is being drilled. The driller's depth and all logs are measured from the same datum and should agree. The driller's depth and recorded log depths are referred to as the *measured depth (MD)* of the well (Fig. 4–1).

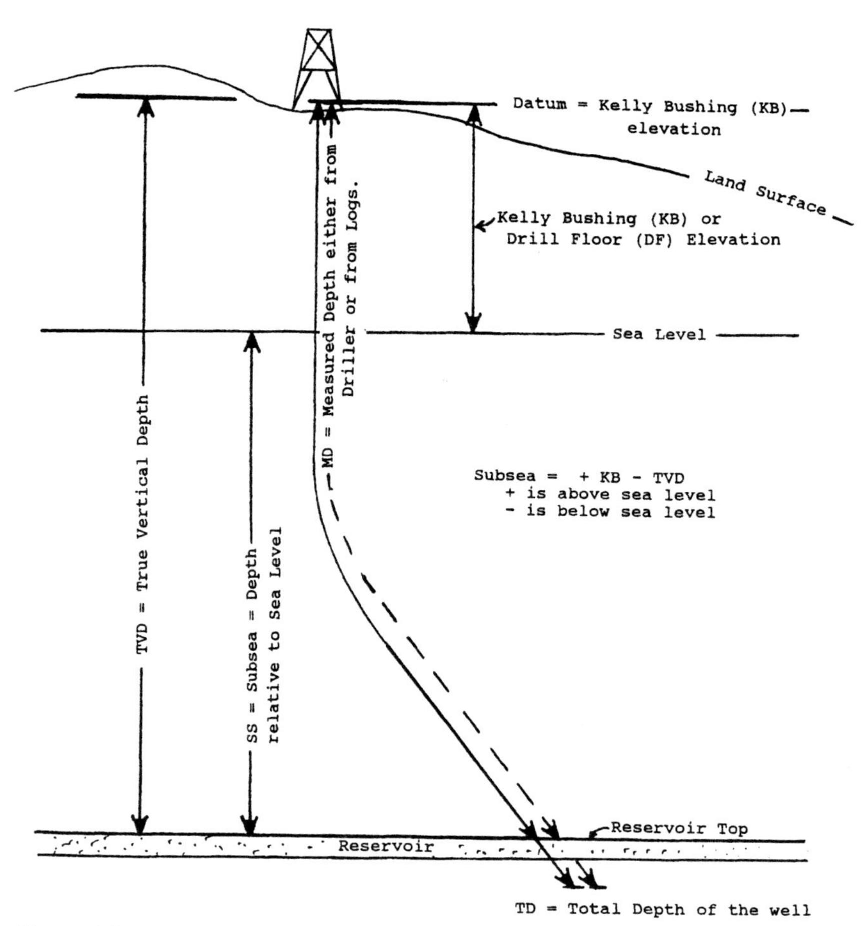

Figure 4.1
Cross section showing the difference between Measured Depth (MD), True Vertical Depth (TVD), and Subsea (SS) elevation.

In deviated holes, the *true vertical depth (TVD)* is the vertical component of the measured depth (Fig. 4–1), and is always less than the measured depth. For a straight hole, TVD and MD are the same. TVD is important because hydraulic gradients (see Chapter 6) caused by fluid in the hole or in the earth operate as a function of TVD, not MD.

All structure contour maps today are made as elevation contours relative to sea level. The petroleum industry's term for these elevations is *subsea (SS),* though the term is confusing because not all subsea elevations are below sea level. Subsea elevations are given by:

Subsea (SS) elevation = + Datum elevation – TVD

For example, if the Kelly bushing datum is +8,000 feet and the TVD to the top of a sandstone is 6,000 feet, then the subsea elevation of the top of the sandstone is +2,000 feet. Except in mountainous terrains, subsea elevations will normally be below sea level and negative.

Because logs are normally digitized today, true vertical depth and/or subsea depth logs can be generated by shrinking the log by the cosine of the hole angle corresponding to the depth on the log. Directional surveys are taken on all deviated holes so that the location of the bottom of the hole and true vertical depths can be calculated.

Caution: TVD logs give correct vertical thicknesses for horizontal beds only. For dipping beds, TVD logs give incorrect stratigraphic and vertical thicknesses. The error is small if the hole angle is small or the dip is small. The error can be very large for holes drilled directionally downdip.

4.2 Scales

In the United States logs are usually run on two scales. Correlation logs are run at scales of either 1 inch = 100 feet, called a *1-inch log*, or 2 inches = 100 feet, called a *2-inch log*. Detailed logs for determination of porosity, water saturation, and pay counting are run at 5 inches = 100 feet, called a *5-inch log*. In many other countries logs are run using metric scales.

4.3 Invasion

While a well is being drilled, fluids from the wellbore enter into permeable formations. Drilling muds are designed such that a *mud filtrate* (Fig. 4–2) forms on the side of the hole and blocks off this permeability so that drilling fluid is retained in the hole. Before the mud filtrate is able to form, fluid from the wellbore displaces or flushes oil, gas, water, or whatever fluid was in the formation away from the wellbore. Beyond the flushed zone, a partially flushed or *transition zone* occurs, and well back in the formation an *uninvaded zone* exists.

A number of different logs, particularly resistivity logs, have different spacings between energy sources and energy receivers. Two important generalizations are:

1. *Short-spaced devices* gather data from very near the wellbore. Short-spaced devices usually show great detail, but the detail is from the flushed or invaded zone.

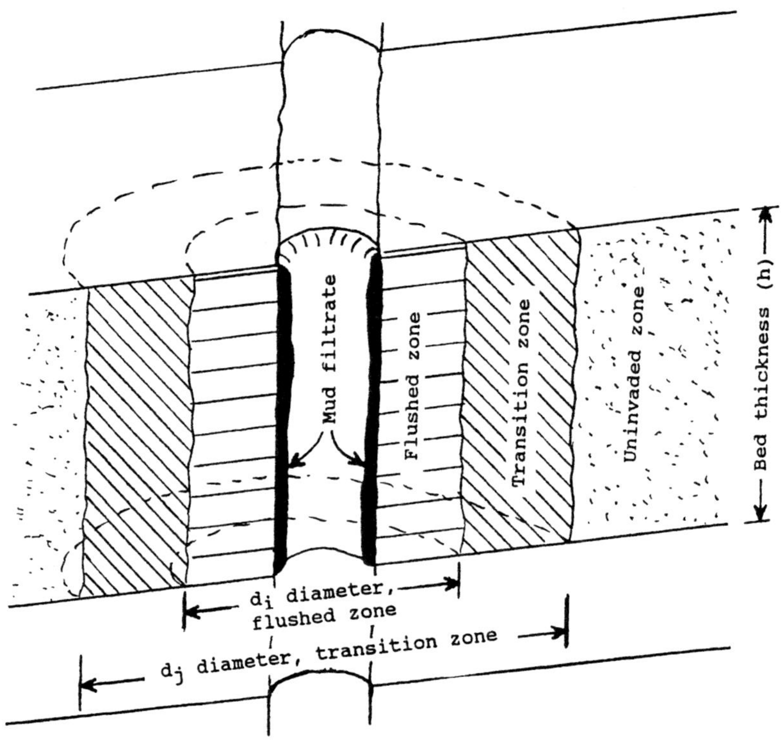

Figure 4.2
Cross section through a wellbore showing bed thickness (h), mud filtrate, diameter of the well bore (d_h), diameter of the flushed or invaded zone (d_i), diameter of the transition zone (d_j), and the uninvaded zone.

2. *Long-spaced devices* gather data from deep in the formation, either from the transition zone or the uninvaded zone. But the data are gathered over long vertical intervals and detail is poor.

Obviously, intermediate-spaced devices read at intermediate penetration depths. Students always ask why there are so many different logs that seem to do the same thing. Each log has strengths and weaknesses, and it is important to understand the trade-offs.

4.4 Correlation Logs: Spontaneous Potential (SP) and Gamma Ray

The two most commonly used correlation logs are the SP (spontaneous potential) and the gamma ray, and they are printed on the lefthand side of most logs (Fig. 4–3). Plotted to the right of the depth scale, most correlation logs are opposed by resistivity which can also be used for correlation, but to a lesser extent.

4.4.1 Spontaneous potential (SP)

The spontaneous potential log (Fig. 4–3) measures the natural battery effect that occurs at the interface where foreign ions from the drilling

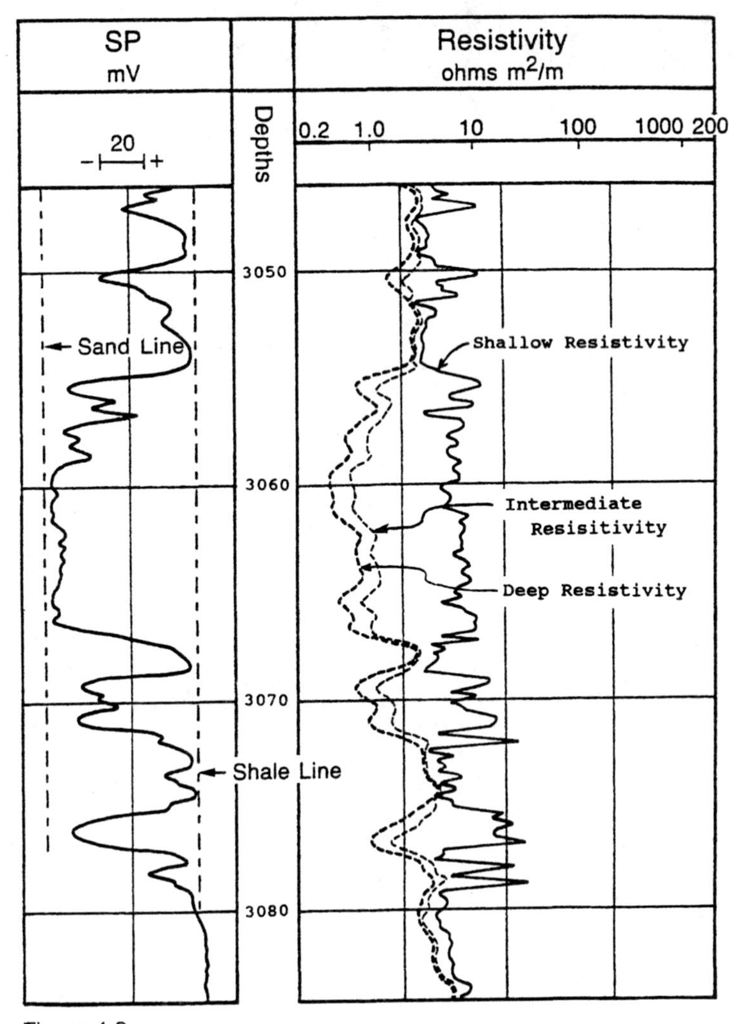

Figure 4.3
Example of Spontaneous Potential (SP) log in a sand-shale sequence. From Schlumberger (1989), reprinted by permission.

fluid have entered into porous zones and are juxtaposed against normal formation fluids in nonporous zones. No battery effect is generated if the drilling fluid and the formation water have the same salinity, and no battery effect is induced across different rock types if both are impermeable and no foreign ions are introduced.

4.4.2 Gamma ray

The gamma ray log acts like a Geiger counter to measure the natural gamma radiation of the rocks as the log is pulled out of the hole. Natural gamma radiation in most rocks is quite low, and as a statistical tool it commonly shows a considerable amount of noise. Gamma radiation is primarily a function of the presence in the rock of radioactive elements such as potassium, uranium, and thorium. Shales contain small amounts of these elements and are radioactively "hot," while sandstones and limestones contain almost no radioactive elements and tend to be "cold."

Both the SP and gamma ray logs are set up such that the logs respond to sandstones and limestones by deflecting (or "kicking") to the left, and shales kick to the right. *Caution:* For the SP, this can be reversed if the salinity of the drilling fluid is greater than the salinity of the formation fluid. This can occur where a saline drilling fluid invades a formation containing fresh water.

4.5 Resistivity Devices

Except for a few metallic sulfides plus graphite and chlorite, dry rock and hydrocarbons are very poor conductors of electricity. The electrical conductivity of a rock is primarily a function of the water that exists in the pore spaces. Electrical resistivity is the opposite (or reciprocal) of conductivity, and is due primarily to three factors:

1. The resistivity of the formation water which is primarily a function of the water salinity
2. Amount of water present
3. Configuration of the pore spaces

Through the years a great deal of attention has been paid to electrical resistivity because it is the principal means by which hydrocarbons can be identified in the subsurface, as well as the principal means by which percent hydrocarbon saturation, one of the fundamental parameters in volumetrics, can be determined. Because hydrocarbon saturation is so important, many different resistivity logs have been developed over the years.

4.5.1 Conventional electrical survey (ES)

One of the oldest and most important of the resistivity combinations is the conventional electrical survey (ES) which consists of a 16-inch normal, a 64-inch normal, and an 18-ft 8-inch lateral device.

The short-spaced device is used for detailed correlations in the invaded zone, and the long-spaced device is used to read deep resistivity beyond the flushed zone. The intermediate device obviously reads between these depths. By comparing readings from these devices against each other, it is possible to determine the depth of penetration of the drilling fluid, and thus gain information on the permeability of the rock. The terms “normal” and “lateral” refer to specific electrode arrangements that give slightly different responses in different situations. The details of these logs are beyond the scope of this discussion, but briefly, these logs have two weaknesses. The normal electrode arrangement is not well focused and is difficult to use quantitatively in thin beds. The lateral arrangement gives asymmetrical results that are also difficult to use in thin beds. Asymmetry refers to the logs’ tendency to read too high near the top of a porous zone, and too low near the base of the zone.

4.5.2 Focused resistivity devices

To overcome these weaknesses, a number of different logs have been developed, including focused laterologs and spherically focused logs. The focused laterologs and the spherically focused log employ bucking electrodes to focus the currents near the wellbore. Laterologs focus the currents in horizontal sheets around the central electrodes, while spherically focused logs focus currents in a sphere around the central electrodes. The spherically focused log is a good log for shallow investigations, and the laterologs are used primarily for deeper investigations. The laterologs still give asymmetrical results, and all need corrections for specific hole conditions.

4.5.3 Induction logs

Induction logging tools were originally developed to be used in nonconductive (oil-base) muds or air-drilled holes, but they are now used in many different hole conditions. The induction log induces currents in the formation through the use of transmitter coils and induced magnetic fields in the formation. These can also be focused through the use of bucking coils, and the induction spherically focused log, in combination with other resistivity devices (commonly the dual induction log), is one of the most frequently used logs today.

4.5.4 Phaser induction logs

Phaser logs not only record currents that are in phase with the transmitter currents, but also currents that are out of phase (usually by a quarter of a wavelength). The dual induction spherically focused log can theoretically resolve thin beds down to a thickness of two feet, and it can be corrected for borehole effect and asymmetric effects.

4.6 Sandstone-Shale Sequences

A summary of the principal formation evaluation logs used today is given in Figure 4–4. To identify hydrocarbons in sandstone-shale sequences, the most important rule is "thumbs out." That is, look for a left kick (deflection) on the SP or gamma ray to find the sandstones and, opposite the sandstones, look for a right kick on the resistivity log. A right-kicking resistivity means that the rock is electrically resistive and that either:

- the rock has small or no pore spaces (limestone or cemented sandstone), or
- the fluid in the pore spaces is highly resistive or will not conduct electricity well. Normally, this means hydrocarbons.

Thus, if the SP or gamma ray kicks left and the resistive curves kick right (thumbs out) in a sand-shale sequence, this usually means hydrocarbons. But—not always. It could mean tight (low-porosity) limestone, tight sandstone, or possibly fresh water in the pore spaces. The porosity logs are used to distinguish between these various combinations.

4.7 Porosity Logs

The most commonly used porosity logs are the sonic or acoustic log, the formation density log, and various types of neutron logs.

4.7.1 Sonic log

The sonic or *acoustic log* is relatively simple, and can be run in combination with various other logs such as the SP and resistivity. An ISF-sonic

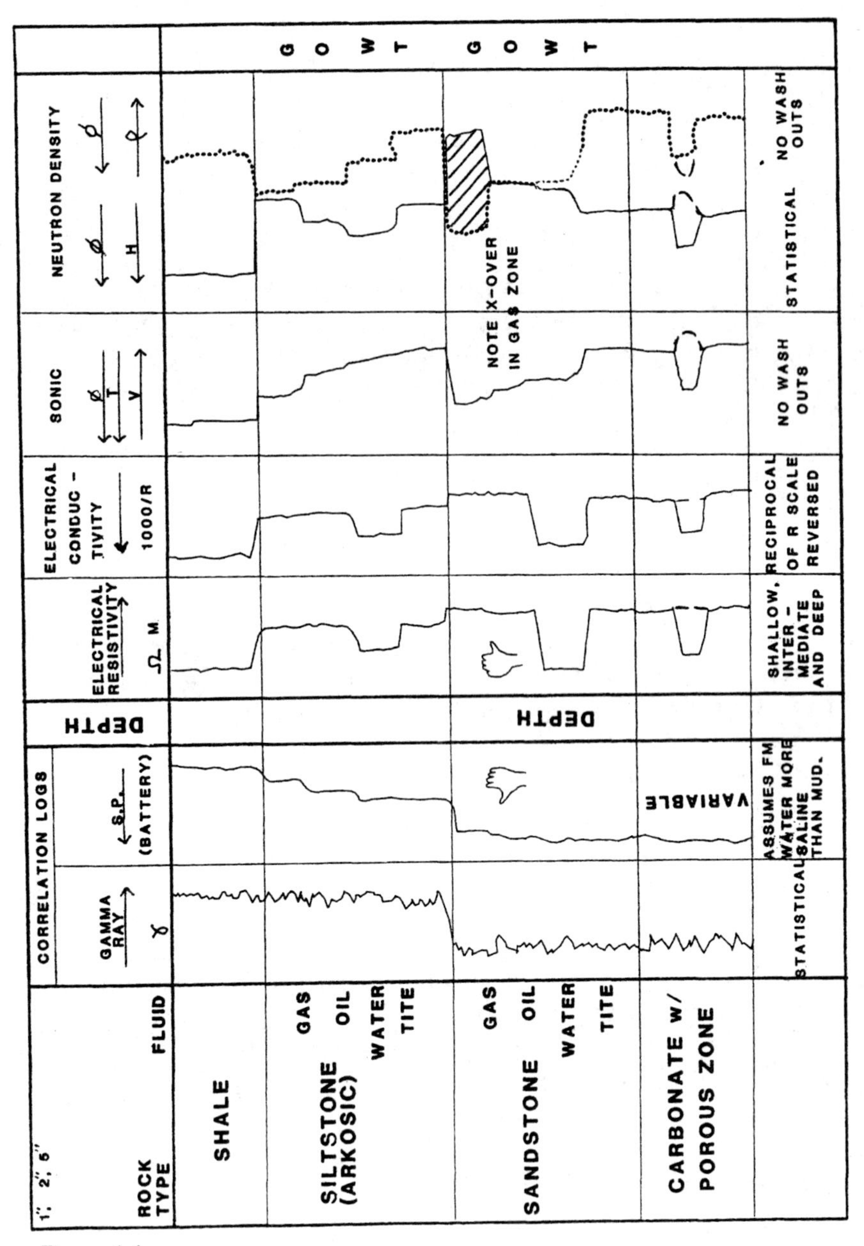

Figure 4.4

Overview of correlation and porosity log responses in different rock types. Thumbs out indicates hydrocarbon-bearing zones.

combination (induction, spherically focused, in combination with a sonic log) is commonly one of the first logs run in both exploratory and development wells. It is one of the best combinations for the money.

The sonic log measures interval transit times (Δt) of acoustic waves (sound waves) over discrete distances. The interval transit time is a function of both lithology and porosity. If lithology is known, the porosity is fairly easy to deduce, as long as the hole is not badly washed out. In badly washed out zones, or in gas zones (which have a high acoustic impedance), cycle skipping and erroneous results may occur. Most sonic logs today are calibrated to specific reservoir rock types, such as sandstone, so that porosity for that rock type may be read directly from the log.

The sonic log is extremely important to the reflection seismologist because correct velocity information is critical for proper stacking, migration, amplitude studies, and conversion of time to depth on seismic lines (see Chapter 14).

4.7.2 Density log

The *density log* is a skid-mounted device that emits medium-energy electrons and measures returning gamma rays. The gamma rays are generated through Compton scattering as a result of collisions with electrons in the formation. Electron density is a function of rock matrix material, porosity, and density of the enclosed fluids. The tool is constructed such that it compensates for hole diameter, but it does not work well in badly washed out holes. This is a very accurate and reliable tool except in washed out holes.

The density log is also important to the reflection seismologist because the amount of acoustic energy reflected at an interface in the subsurface is principally a function of density and velocity change across that interface (see Chapter 14).

4.7.3 Neutron log

Several different types of neutron logs are available. Most emit high-energy neutrons and measure some sort of returning radiation that is a function of the elemental hydrogen content of the reservoir. They are relatively inexpensive, are not particularly affected by washouts or chlorine in the mud or formation waters, and can be run inside cased holes. They are especially effective when run in combination with other porosity tools such as the formation density log. In Kansas and Oklahoma, where the most important logging considerations are cost and identification of porous zones, the neutron log is commonly run in combination with the gamma ray.

4.7.4 Density-Neutron combination

All porosity logs are designed to measure porosity in the presence of water. All of the porosity logs tend to overreact to a large degree in the presence of gas, and to a small degree in the presence of oil. This over response can be used effectively to distinguish between gas and oil.

One of the best ways to identify gas in a formation is to run the density and neutron logs simultaneously. The logs should be calibrated to read correct porosities in water-bearing formations. In gas-bearing zones, the over-response is such that the two curves cross each other (Fig. 4–4). This crossover of the two logs is one of the best ways to distinguish gas from oil in a hydrocarbon-bearing zone. Be aware that the over-response is diminished in low-porosity zones such as in siltstones.

Porosity in a gas-bearing zone can be estimated using the following equation:

$$\text{porosity}_{\text{gas zone}} = \left[\left(\phi_d^2 + \phi_n^2\right)/2\right]^{\frac{1}{2}}$$

where: ϕ_d = porosity from the density log.
ϕ_n = porosity from the neutron log.

4.7.5 Distinguishing between gas and oil

Gas in a reservoir can usually be identified by the over-response of porosity logs (above) or by cycle skipping on the sonic log.

Oil can be identified by three methods:

1. A nongas, hydrocarbon response in a porous zone (above).
2. Fluorescence in samples. Most, but not all, oils fluoresce when exposed to ultraviolet light. Cuttings or cores saturated with oil will usually fluoresce gold, yellow, or sometimes blue when exposed to UV light.
3. Oil derived from cuttings or core chips, when exposed to organic solvents such as trichloroethane, will cause the solvent to discolor to gold or yellow. When this happens, the solvent is said to have an “oil cut.” A strong cut is one of the best indications of oil. *Caution:* If diesel fuel or an organic based lubricant, such as

"Black Magic," have been introduced into the wellbore (usually to free stuck drill pipe), all cuttings may fluoresce or give an oil cut.

4.8 Cased Hole Logs

One of the most important functions of the development geologist is to look for undrained hydrocarbon zones in existing wells. Logs that are able to see through casing are obviously very important for this type of analysis. Through-casing logs can also be used effectively in highly deviated holes, holes where open hole tools cannot be used easily because they will not fall freely in the uncased wellbore. The three through-casing logs are the gamma ray (above), the neutron log (above) and the pulsed neutron log.

4.8.1 Pulsed neutron logs

Pulsed neutron logs, such as the thermal neutron decay time log (TDT), emit pulses of neutrons, and measure returning radioactivity as a function of time. The pulsed neutron log response is similar to that of a resistivity log and, used in combination with the gamma ray and neutron logs, can be used almost as effectively as open hole logs.

4.9 Archie Equation and Water Saturation

In 1942 Gus Archie of Shell Oil Company published the following equation, which is now known as the Archie equation for water saturation.

$$S_w^n = \frac{F \cdot R_w}{R_t} \qquad \text{(Equation 4–1)}$$

Two additional equations, which have now been modified, are:

	Original Archie		*Modified*	
For carbonates:	$F = 1/\phi^2$	or	$1/\phi^{2.2 \text{ to } 2.5}$	(Equation 4–2c)
For sandstones:	$F = 1/\phi^2$	or	$.62/\phi^2$ or $.81/\phi^2$	(Equation 4–2s)
	$R_0 = F \cdot R_w$			(Equation 4–3)

where:

S_w = water saturation expressed as decimal

n = power, based on rock type:

1.8 for sandstones

2.0 for limestones & dolomites

2.5 for chalks

F = Formation factor and is a function of rock type and porosity as demonstrated in Equation 4–2.

R_w = electrical resistivity of the connate (formation) water; is a function of salinity.

R_t = the true electrical resistivity of the hydrocarbon bearing formation. True means beyond the invaded zone and must be read from a deep reading device, or adjusted from a shallow reading device.

ϕ = porosity expressed as a decimal.

R_0 = R_{zero} = electrical resistivity of the hydrocarbon-bearing formation where it contains zero hydrocarbon saturation or 100 percent water saturation.

R_0 is *not* the resistivity of oil and it is *not* the resistivity of the water. Resistivity of the water is R_w, and resistivity of oil is not measured.

Note several important features about the equations. Of particular significance is that R_0 and R_t *must be read from different locations.* It is not possible to read the electrical resistivity of the hydrocarbon-bearing zone at the same location as the electrical resistivity of the water-bearing zone. If the rock type or the porosity or salinity of the formation waters change between the two locations, adjustments must be made, or the equations do not hold.

Another important point is that if Equation 4–3 is plugged into Equation 4–1, and we assume that $n = 2$, a shorthand Archie equation (Equation 4–4) can be developed.

$$S_w = (R_0/R_t)^{\frac{1}{2}} \qquad \text{(Equation 4–4)}$$

This is a very handy equation, but it is somewhat dangerous because it is too simple in assuming that:

- $n = 2$, which may or may not be true,
- The porosity is the same at the locations where both R_0 and R_t were read,
- R_w is the same at the locations where both R_0 and R_t were read, and
- n has not changed between the two reading locations.

The ideal situation is shown in Figure 4–5. Note that the upper part of the sandstone is hydrocarbon-bearing, and the lower part is wet,

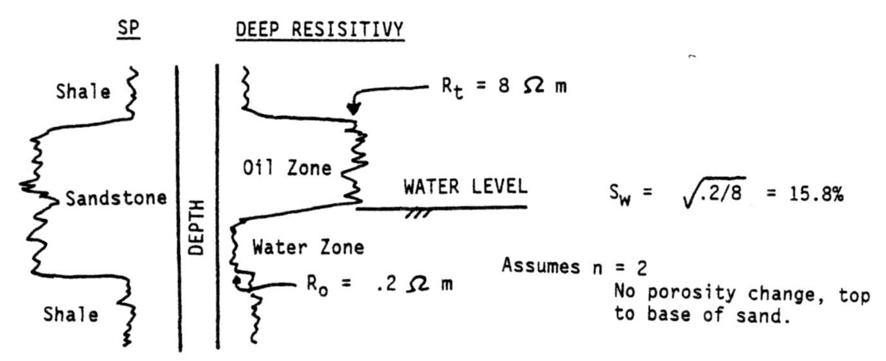

Figure 4.5
Ideal situation for use of the shorthand Archie equation.

with an oil-water contact in the middle. Using Equation 4–4, a reasonably accurate estimate of the water saturation is shown.

A more common situation is shown in Figure 4–6. The shorthand notation is acceptable if porosity and R_w (function of salinity) are the same in the two sandstones. It is usually reasonable to assume that R_w will be similar for two nearby sandstones if:

- they are not separated by a great vertical distance,
- there is no intervening fault,
- there is no intervening unconformity, and
- if there are no known salinity changes in the area.

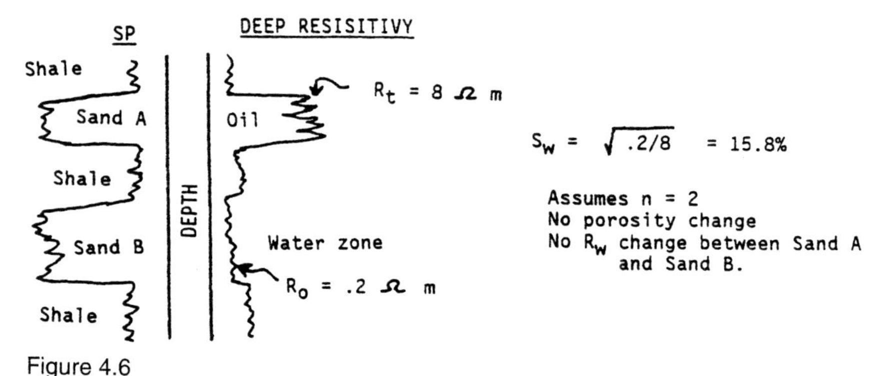

Figure 4.6
Usually acceptable situation for use of the shorthand Archie equation. This situation is not acceptable if porosity or water salinity is different in the two sandstones.

Another common situation is shown in Figure 4–7. Because the porosity in the two zones is different, the shorthand equation cannot be used, but the full set of Archie equations can be used.

If the shorthand Archie equation is used, a water saturation of 40 percent is calculated. However, we have just stated that the shorthand equation cannot be used where the porosities of the two sandstones are different. What can be done?

By using sandstone B, plus Equations 4–2 and 4–3, an R_w can be calculated for sandstone B. If we assume that the connate water resistivity (R_w) is the same for both sandstones A and B, then an F and R_0 can be calculated for sandstone A. With an R_0 and an R_t, we can then estimate S_w for sandstone A as follows:

1. Using sandstone B and Equation 4–2s:

$$F_B = .62/\phi^2 = .62/.24^2 = 10.76$$

2. Using sandstone B and Equation 4–3:

$$R_w = R_0 / F = .8 / 10.76 = .074$$

3. Using sandstone A and Equation 4–2s:

$$F_A = .62/\phi^2 = .62/.3^2 = 6.89$$

4. Using sandstone A and Equation 4–3:

$$R_0 = F \cdot R_w = 6.89 \cdot .074 = .510$$

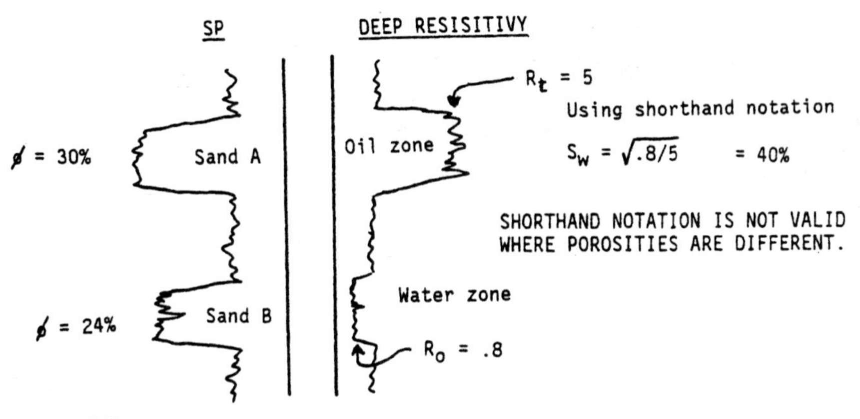

Figure 4.7
Shorthand Archie equation is not valid where different porosities or water salinities are encountered in different sands. It is particularly not valid if Sand A contains fresh water.

5. Using sandstone A and Equation 4–4:

$$S_w = R_0/R_t = .51/5 = .32$$

Thus, a water saturation of 32 percent rather than 40 percent is calculated. Thus, by using the Archie equations, a correction can be made for one of the three parameters (rock type, porosity, or R_w) by assuming that the other two have not changed between the locations where R_0 and R_t are read.

4.10 Electromagnetic Propagation Logs

An important new suite of logs that has been developed over the last ten years is the electromagnetic propagation log, sometimes referred to as a dielectric log or phaser log. These logs are still in the development phase, but they appear to have great potential for identification of different clay types and water saturation.

The dielectric permittivity of most reservoir material is primarily a function of water-filled porosity. The dielectric permittivity of water is relatively independent of salinity and ranges from about 25 to 30. The permittivity of most other reservoir material is less than 8. Thus, the log can be used effectively for identification of water saturation almost completely independent of water salinity. The log run by itself does not distinguish between connate water, mud filtrate, and bound water associated with shales.

However, a shallow reading device used in combination with a deep reading device can be used to attempt to identify these different water types. The log works on the principle that electromagnetic waves are attenuated and experience phase shifts as a function of dielectric properties of the rock and frequency of the waves. High-frequency waves are used for shallow investigations, and low-frequency waves are used for deeper investigations. The author's impression of the device is that the shallow reading device works well, but the deep reading device is dependent on so many variables that its reliability is subject to many different correction factors, and its reliability is poor.

4.11 Oil-Water Contact

Throughout this discussion, it has been assumed that the oil-water contact is a planar surface at the base of oil. In fact, the oil water contact is always a transition zone. In clean sandstones or vuggy limestones, the transition zone may occur over a vertical thickness as small as several inches. But, in reservoir rocks that contain very small pore throats, such as siltstones or micritic limestones, the transition zone may be spread over a vertical thickness of as much as 50 feet. Thick transition zones are common in low-porosity rocks because capillary forces in such rocks are commonly considerably stronger than gravitational segregation forces.

4.12 Digitized Well Logs

Within the last 20 years, the availability of digitizers has become important to the development geologist. Old electric logs which were originally plotted on a linear scale may now be digitized, calibrated, and replotted on a logarithmic scale for comparison against newer logs. A comparison of newer logs against the old, now calibrated logs, allows the development geologist to look for changes that have (or have not) occurred since the older logs were run. The ability to do this is extremely important in the reevaluation of old fields because it allows the development geologist to determine which zones have been depleted and which have not.

4.13 Dipmeter

The dipmeter is one of the most important tools to the development geologist. It has been around in various forms for a long time. Recent modifications, and its use in combination with other logs have made it into a potentially powerful tool. Only a brief discussion of the dipmeter follows, as excellent documentation of the tool is available through logging companies.

4.13.1 Older dipmeters

The oldest dipmeters consisted of three microresistivity devices mounted on pads. The geologist basically correlated the microresistivity devices by hand and worked three-point problems to solve for the orientation of the beds.

In deviated holes, it was found that one of the three arms commonly did not contact the side of the hole, and early on it was decided that four arms were needed. The four-arm dipmeter consists of four microresistivity devices mounted on pads, and data is recorded digitally. Through a process known as cross correlation, digitized data are then correlated among the four arms by the computer, and the three best correlations are used to solve three-point problems for the dip of the bed. Because all data are recorded digitally, corrections can be made for orientation of the tool, asymmetry of the hole, and other problems. The data can then be presented in a number of fashions, the most common being the tadpole plot (Fig. 4–8), where poorer correlations are symbolized by open tadpoles, and better correlations are symbolized by filled tadpoles.

4.13.2 Cross correlation

The cross-correlation process is an important process used in modern dipmeter interpretations. It is important that the geologist have an idea as to how it works. Digitized data from two microresistivity devices are correlated mathematically over an *interval length,* by increments of *step lengths,* and with a *search angle.* In the cross-correlation procedure the computer sums the absolute mathematical difference between data points from the two devices over the interval length. The computer then offsets the data by one step length and sums again. It then offsets the data by another step length and compares again. Where the correlation is best, that is, where the two sets of data overlay each other almost perfectly, the sum of the absolute difference between data sets will approach zero. Thus, the computer is able to pick the best correlation over that interval length by looking for the minimum absolute difference location over the interval length. Clearly, if the best fit is between steps, it may never find a good correlation. And if the data sets are regular cycles (such as a sine wave), the computer may pick the wrong peaks as the best correlation. Further, if the best correlation is outside the interval length or the search angle, the computer will pick a best fit and will give a false correlation. The method can work very well, but the user needs to be aware that there is always scatter and potential problems in the data.

Caution: Dipmeters do not work well in badly washed out or irregularly shaped holes or in highly deviated holes drilled directionally downdip.

The second case results in the hole penetrating the beds at a very high angle. If the search angle or correlation length are not large enough, the computer will make incorrect correlations that are commonly just below the search angle, and this is one way to identify a bad dipmeter. The data can be recorrelated using higher interval lengths and larger search angles, but in doing so, more false correlations are likely to appear.

4.13.3 Modern dipmeters

The modern dipmeter is a four-arm dipmeter with at least two microresistivity devices on each arm. Cross correlation of the double resistivity

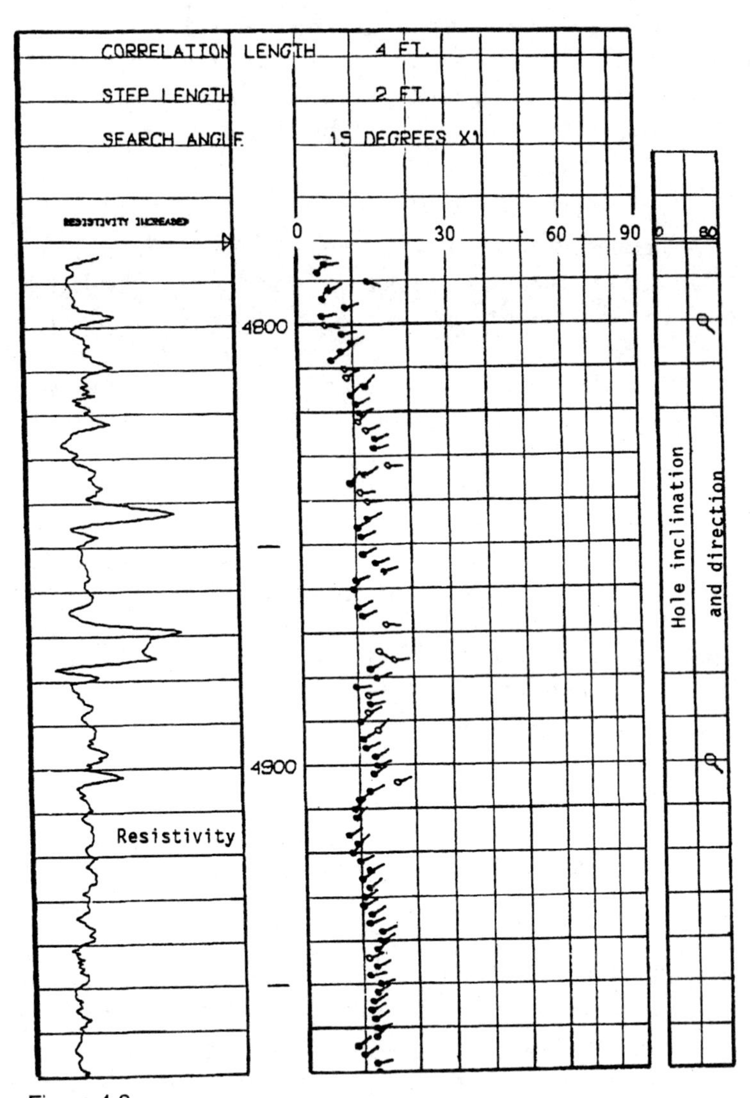

Figure 4.8
Typical tadpole plot for an offshore well. Beds are dipping consistently to the northeast at 10 to 12 degrees. The hole is deviated to the southeast at about 56 degrees. Solid tadpoles represent "good" cross correlations. Open tadpoles are "poor" correlations.

devices allows for confirmation that both devices are recording properly and increases confidence in all correlations. In some cases the dual devices can be used for identification of vertical fractures. Visual images derived from all eight devices can be used to give a visual image of the well bore that is similar to some of the modern microscanning devices. In addition, the newer devices have better pad-mounted suspension devices and accelerometers, which give much more accurate results in deviated and irregularly shaped holes and when the tool is tending to yo-yo in the hole.

4.14 Vertical Sequences

Vertical dip sequences can be used very effectively today to identify various structural and stratigraphic relationships. In combination with lithology logs, they can be used very effectively in some cases for identification of sedimentary environments.

Schlumberger defines a *blue* dipmeter pattern as an upwards increase (IU) in dip and a *red* dipmeter pattern as an upwards decrease (DU) in dip. The following is a summary of how these patterns should theoretically appear in a wellbore. These generalizations work with varying degrees of success.

1. *Faults* can be red (DU) or blue (IU) (Fig. 4–9).
2. *Unconformities* are normally red (DU).
3. *Current bedding* is normally blue (IU) (Fig. 4–10), but can be chaotic within a sandstone unit. This is largely a function of the thickness of the principal bedding.
4. *Channel sandstones* are normally red (DU) (Fig. 4–10) with principal dips towards the axis of the channel. Smaller-scale blue patterns (IU) caused by cross bedding are likely within an overall red (DU) pattern. Channel sandstone should be fining upward on gamma ray or SP.
5. *Barrier island sandstones* should be red (DU) (Fig. 4–10) with dips away from the axis of the barrier island. Some smaller-scale blue (IU) patterns caused by cross bedding are likely to be imbedded within the overall red (DU) pattern. Barrier island sandstones should coarsen upward on gamma ray or SP.
6. *Reef* cores will normally show chaotic dips. Drape over the crest will show a red pattern (DU) that dips away from the crest of the reef (Fig. 4–10). If the reef has foundered, underlying beds will be blue (IU) and will dip towards the reef.

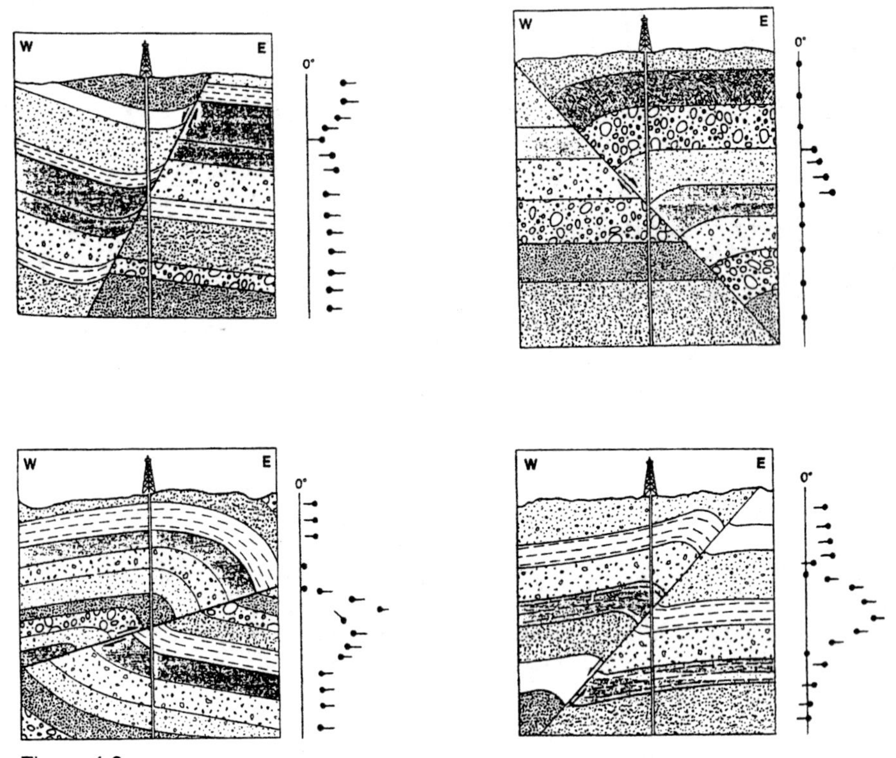

Figure 4.9
Dipmeter patterns associated with various fault types. Normal faults with normal and reverse drag at the top, and thrust fault and reverse fault below. Note both red and blue patterns. From Schlumberger (1986), reprinted by permission.

4.15 Some Commonly Asked Questions

1. If maps are made relative to sea level, how do you get subsea elevations in a deviated hole?

Answer:

Subsea elevation = Datum – TVD elevations.

True vertical depths come from a number of different types of directional surveys that are run while the well is being drilled and are very accurate today. Today gyroscopes and sensors are located in a drill collar behind the drill bit, and data is transmitted to the surface by sonic pulses in the mud. This technology is relatively new and is extremely important for horizontally drilled wells. Other techniques employ wireline devices (single shot and multishot) that are dropped into the hole and take snapshot photos of a bubble-mounted compass.

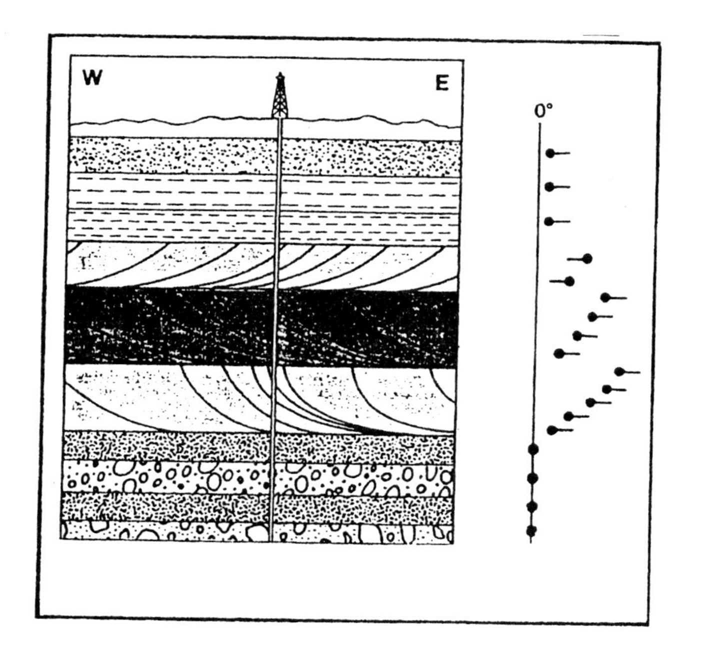

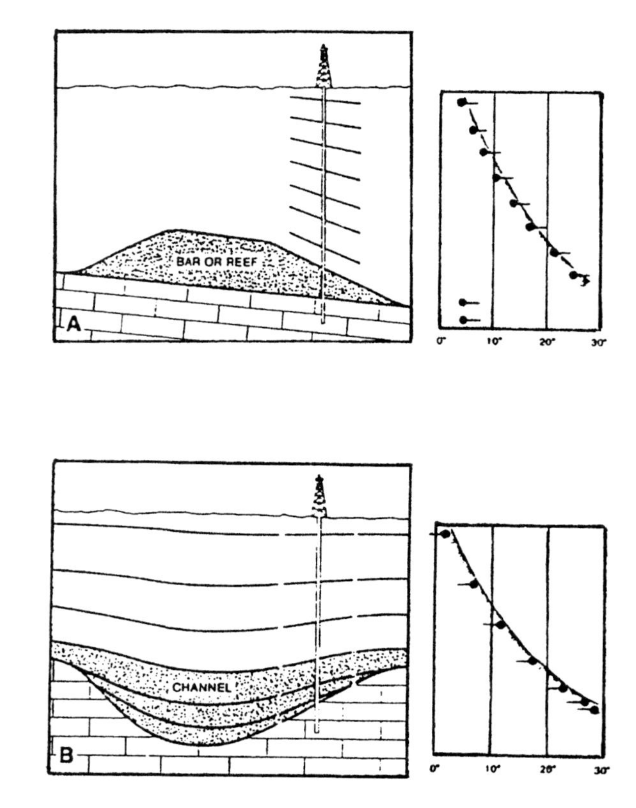

Figure 4.10
Dipmeter patterns for large-scale cross bedding (blue pattern, on left), drape over a reef or barrier island (red pattern, upper right), and dip into a channel sand (red pattern, lower right). Courtesy of Schlumberger (1986), reprinted by permission.

2. If the combination neutron-density log is good for telling oil from gas, how do you tell oil from water?

Answer:

Don't forget resistivity.

3. R_0 is resistivity of oil. True or false?

Answer:

False.

4. Can fractures be identified in logs?

Answer:

There are a number of new microscanning devices that give a virtual picture of the walls of a wellbore. Typically they must be run very slowly, and they are quite expensive, but some new devices are very impressive.

CHAPTER 5

Drill Time Logs, Mud Logs, Cuttings, and Cores

While a well is being drilled, a great deal of information is usually recorded by some sort of continuous recording device (e.g., a geolograph) or by an individual or group on the drilling rig. Information that is almost always recorded include the following:

1. Drill time records which can then be plotted as a drill time log (Fig. 5–1)
2. Cuttings, which are simply small rock chips created by the drill bit
3. Mud logging devices, including:
   (a) Mud pit volume indicator. If the mud level starts to rise, this is the first indication that the well is attempting to kick and that a blowout may result. It means that fluid is entering the wellbore somewhere.
   (b) Gas detection system. When an unusual amount of natural gas comes out of the mud, a "gas show" is recorded.
   (c) Gas chromatograph. A gas chromatograph gives an analysis of the gas generated. If long chain hydrocarbons are detected or if small amounts of crude oil occur in the mud, then an "oil show" is recorded.
   (d) Mud weight in pounds per gallon (ppg)
   (e) Weight on the drill bit

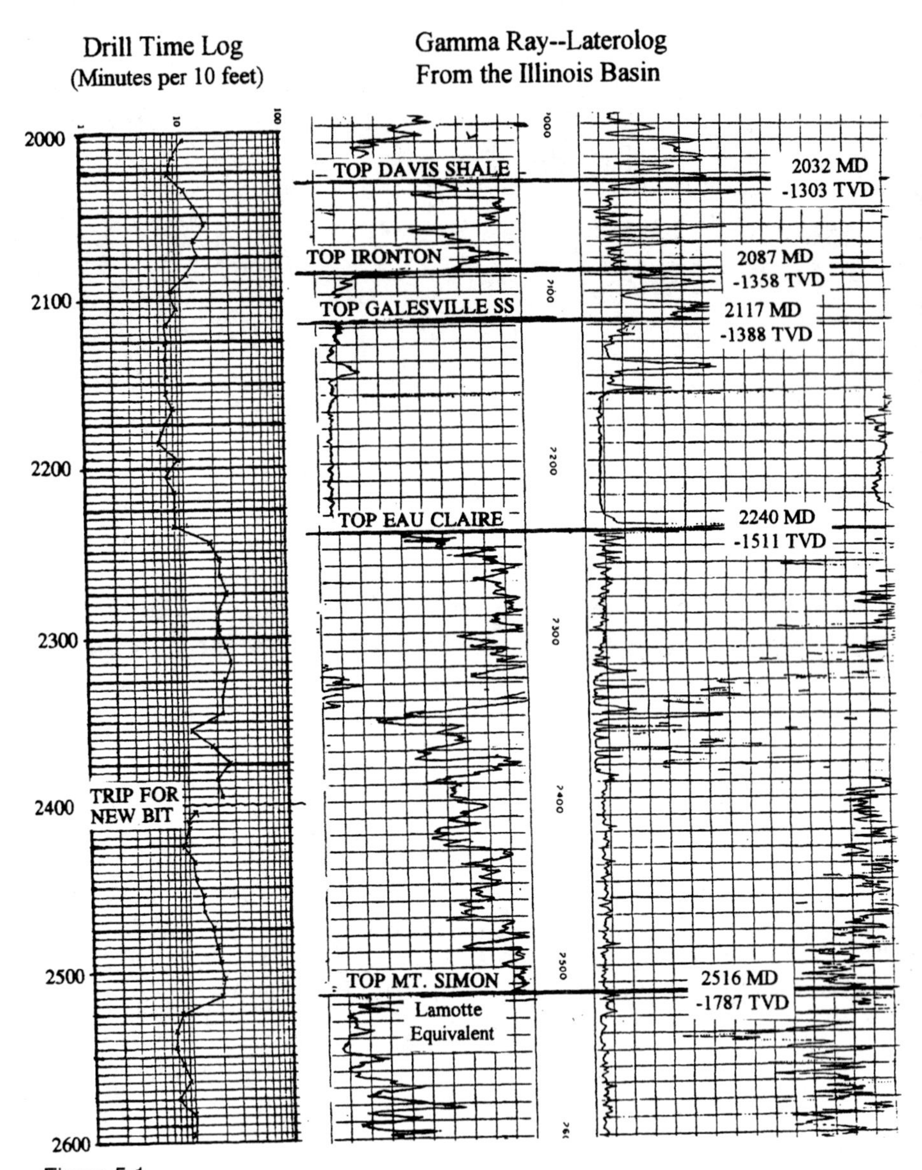

Figure 5.1
Comparison of drill time log to electric log in the Illinois basin. Note the logarithmic scale for the drill time log.

(f) Direction and location of the drill bit

(g) Other mud properties, including viscosity, pH, fluid loss, electrical resistivity, and more

When an extremely important horizon is reached, a core may be taken. To do this, the drill pipe is pulled from the hole, the regular drill bit is taken off, and a core barrel is attached. A core barrel is a hollow drill pipe with a diamond-studded grinding base and an internal device that keeps the core from falling out when the core barrel is pulled from the hole. Because the core barrel must grind its way through the rock, drilling with a core barrel is very slow and doesn't always work. Missing section occurs in all cores. The running of continuous core is very slow (relative to drilling) and is thus expensive. Continuous cores are fairly rare in the petroleum industry because they are so expensive. They are run today only in cases where the rock properties are extremely critical or where modern electric logs are not capable of obtaining the desired information.

Of particular interest to the geologist are drill time logs, cuttings, and cores. We will next examine these methods.

5.1 Drill Time Log

The first indication of what rock type the drill is penetrating is the speed with which the bit penetrates. Remarkably, if a record of the rate of penetration is kept, one can commonly develop a log that is almost as good for correlation as a standard electric logs (Fig. 5–1). If the hole is lost for any reason, the drill time log may be the only record of what that well penetrated. In order for the drill time log to have meaning, however, one must have some experience in the area where the well is being drilled. For example, in the Gulf Coast sandstones drill fast and shales drill slowly. In hard rock areas the reverse may be true.

Drill time records are normally kept by the driller (main person on the drill floor also runs the draw works). The driller keeps a record of how long it takes to drill every 10 feet. The drill time log is a plot of drill time versus depth, and it may be done either on a linear scale or a log scale. In the author's experience a good scale is a two-cycle log scale (Fig. 5–1) from 1 minute/10 feet to 100 minutes/10 feet, with one minute on the left side of the scale, such that fast sections (sandstones) kick to the left and slow sections (shales) kick to the right as on an SP or gamma ray log. *Cautions:* The drill time log will be inaccurate due to several factors:

1. If the bit wears out; a dull bit will result in slow drilling.

2. If the bit is changed; a new bit will speed up dramatically, and a bit change should be recorded on the drill time log.
3. If additional weight is applied to the drill pipe and hence the drill bit; rate of penetration will increase.
4. If the mud weight is increased; the rate of penetration will decrease because the drill pipe is, in effect, being floated and weight on the drill bit is decreased.
5. If you drill into high pressures without knowing it, the drilling rate may increase subtly or dramatically. This is because the effective mud weight will be decreased by the addition of formation gas to the mud.
6. Directional drilling, a change in bit type, and many other things can cause penetration rate changes.

5.2 Cuttings and Cores

Both exploration and development geologists commonly like to be at the drill site when important target horizons are penetrated. First impressions of both cuttings and cores coming to the surface can be extremely important in the description and understanding of important reservoir rocks. The following section is a brief discussion of some of the procedures used in describing cuttings and cores, a critical job for both development and exploration geologists.

From cuttings and cores, geologists usually make a strip log. This usually consists of a vertical lithologic profile as a function of depth, with descriptions to the side. Standard lithologic symbols are used, and the strip log is commonly color coded (Fig. 5–2). Descriptions should be in standard format as described below. The strip logs can then be compared to electric logs, drill time logs, or vertical plots of any type.

Cuttings present some special problems. Cuttings are small chips of rock carried up from the bit in the drilling mud. As such, it takes a certain amount of time, known as *lag time,* for the chips to reach the surface after the bit has penetrated the horizon. Further, turbulence in the drilling mud mixes cuttings from overlying horizons. Thus, in attempting to identify the rock that originates from a particular drilling break (change in drilling speed), geologists look for the first appearance of that rock type after the calculated lag time has occurred.

Lag time is a function primarily of depth, mud volume, and pump speed, and varies depending on drilling conditions. Lag time can be as little as one minute, or as much as 60 minutes. Commonly it is 15 to 45

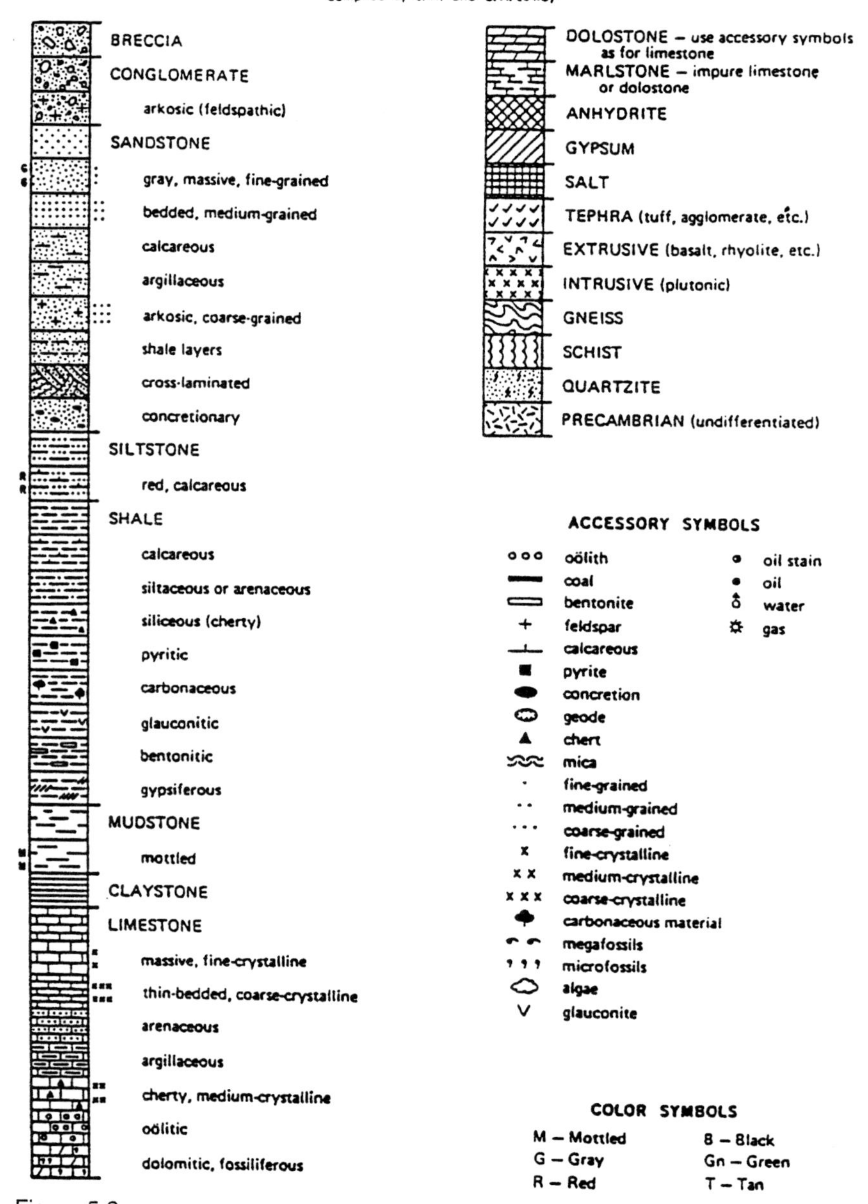

Figure 5.2

Graphic symbols commonly used for plotting well logs. Standard colors: carbonates = blue; sandstone = yellow; coal = black; evaporites = orange; shale = red, green, or grey, depending on color of shale; gas = red; oil = green; water = blue. From LeRoy and LeRoy (1977), reprinted by permission.

minutes. If the drilling break is critical, the geologist may want to ask the drilling foreman (tool pusher) or driller to stop drilling and circulate "bottoms up" so that cuttings from the drilling break can be examined. The tool pusher knows the volume of mud in the hole, and the pump speed, and the geologist should be able to develop a pretty good idea of the lag time simply by talking with the tool pusher. Communication is again critical. If full circulation time (not lag time) needs to be determined, an accurate way is to inject a tracer (commonly radioactive) in the mud and measure the time until the tracer returns to the surface.

5.3 Cuttings and Cores Out of the Hole

Cuttings and cores, coming out of the hole, will be a muddy mess, but the first observations can be extremely important when looking for oil and gas, particularly with cores. Cuttings and cores that have just come out of the hole have just been relieved of their confining pressure, so this is an important time to look for oil bleeding from the core or oil in the cuttings. Cores should initially be wiped clean with a rag so that it is clear exactly where the oil might be bleeding from. If the core is washed down with a fire hose, it may not be possible to identify the exact horizons from which the oil was bleeding.

After looking for bleeding horizons, the next step is to label the core and the boxes very carefully. *It is critical that cores and core boxes be labelled immediately. Depths, tops, and bottoms should be labelled.* It is unbelievably easy to pick up a core and forget which way is up, or forget where it came from. The more labelling that is done early, the fewer problems will result later when it comes to describing the cores. So that cores cannot be mistakenly put in boxes upside down, two different colored vertical lines should be put on all cores such that one color is always on the right and the other color is on the left when the core is oriented right side up. After labelling the cores and the boxes with permanent markers, they can be cleaned up and described.

The same procedure applies for cuttings, although these are typically collected at specific intervals, such as every ten feet, by rig personnel. Do not try to describe them until they are clearly organized and labelled. Generally, cuttings are lightly washed to remove most drilling mud, and wet samples are delivered to the geologist. Typically, the cores and cuttings should be washed, then described, then allowed to dry before being placed in permanent storage.

5.4 Cuttings and Cores Descriptions

Anyone who will be describing cuttings, cores, or stratigraphic sections in the field should develop a standard procedure for those descriptions. Most companies have their own suggested procedures, and the AAPG *Sample Examination Manual* (see the end of this chapter) also has guidelines. It is a good idea to develop a routine procedure, because it is likely that you will not always have the guidelines with you.

One commonly used description procedure is as follows:

1. General rock name—underlined and followed by more detailed rock name
2. Color
3. Texture—including grain size, roundness, and sorting
4. Cement and/or matrix material
5. Fossils and accessories
6. Bedding type—e.g., thin to thick beds or laminations
7. Sedimentary structures—cross bedding, worm burrows
8. Porosity and shows
9. Anything else that is striking or important—e.g., odor. Often, oil-bearing cuttings will have odor due to liberation of gas, and this can be an indicator of live oil.
10. Oil or gas shows such as oil cut

5.4.1 Examples of descriptions

- **Limestone:** Oolitic grainstone, brown, medium to coarsely crystalline, minor brachiopod fragments, some glauconite, thinly laminated with minor cross beds, good intergranular porosity, light amber cut.
- **Sandstone:** Lithic arkose, speckled red, medium grained, poorly sorted, poorly rounded, calcite cement, no fossils, thickly bedded, intergranular porosity, no cut.

Abbreviations are acceptable, but these should be standardized and easy to understand so that other people can understand your descriptions.

5.4.2 Description technique

Rock types General rock types are sandstone, limestone, dolomite, shale, and so forth. More specific rock types may be taken from Figures 5–3 through 5–6. Sandstone examples are given in Figure 5–3,

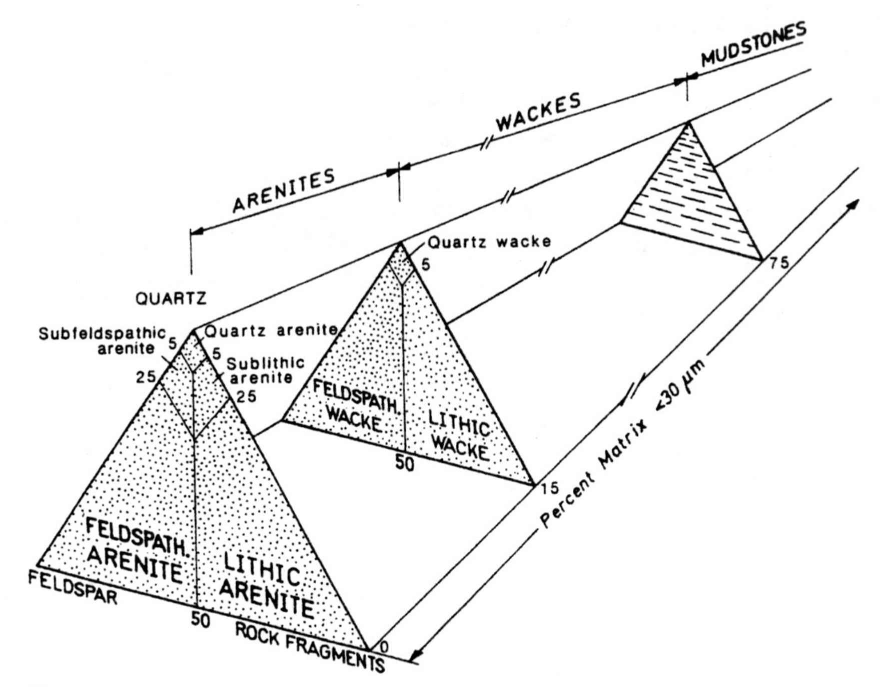

Figure 5.3
Commonly used sandstone classification. Modified from Dott (1964). Reprinted by permission of Society of Economic Paleontologists and Mineralogists.

mudrock examples in Figure 5–4, and limestone examples in Figures 5–5 and 5–6.

Color It is recommended that color be described from wet samples and under 10 power magnification. It is recommended that color be compared to the GSA rock-color chart for standardization. Different eyes and different lighting conditions give different color results.

Silt/ Clay Ratio	Undurated Non Fissile	Indurated Fissile
> 2/3 Silt	Siltstone	Shale
1/3 to 2/3 Silt	Mudstone	Shale
> 2/3 Clay	Claystone	Shale

Figure 5.4
Commonly used classification of mudrocks.

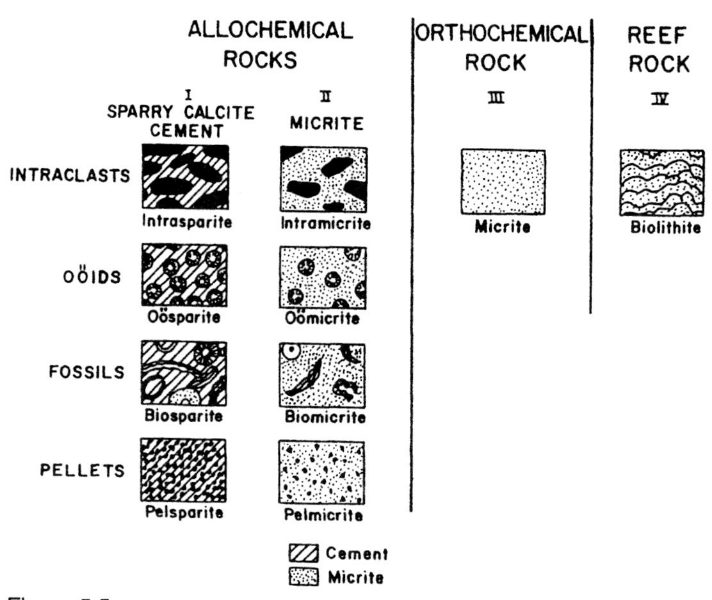

Figure 5.5
Folk's (1959) classification of limestones. Reprinted by permission of AAPG.

Textures Grain size, shape, sphericity, and sorting should all be compared to some sort of visual comparitor. For fine-grained rocks, the comparison should be made under 10 power magnification (hand lens or microscope). For recrystallized rocks, such as dolomite, texture of crystals and crystallinity may be more important than rounding, sphericity, and sorting of the original (pre-recrystallization) grains.

CLASSIFICATION ACCORDING TO DEPOSITIONAL TEXTURE

DEPOSITIONAL TEXTURE				
Original components not bound together during deposition				Original components were bound together during deposition ... as shown by intergrown skeletal matter, lamination contrary to gravity, or sediment-floored cavities that are roofed over by organic or questionably organic matter and are too large to be interstices.
Contains mud (particles of clay and fine silt size)			Lacks mud and is grain supported	
Mud supported		Grain supported		
Less than 10 per cent grains	More than 10 per cent grains			
Mudstone	Wackestone	Packstone	Grainstone	Boundstone

Figure 5.6
Dunham's (1962) classification of limestones. Modified from Swanson (1981), reprinted by permission of AAPG.

Cement and matrix material Recognition of intergranular clay, silt and cements in sandstones can be extremely important, as all of these affect permeability.

Fossils and accessories Fossils and accessory rocks and minerals such as chert, pyrite, carbonaceous material, and glauconite are important for correlation and environmental indicators. All should be noted.

Sedimentary structures Most sedimentary structures cannot be identified in cuttings, but small-scale structures such as cross stratification, climbing ripples, worm burrows, and algal structures can be identified in cores.

Bedding types Bedding types are identified according to the following:

> 1 meter thick	very thickly bedded
30–1000 cm	thickly bedded
10–30 cm	medium bedded
3–10 cm	thinly bedded
1–3 cm	very thinly bedded
.3–1 cm	thick laminations
< 0.3 cm	thin laminations

The word "massive" has no formal meaning. It is used in a relative sense, like "big" or "small."

Porosity This is commonly difficult to determine in hand sample, but obvious vugs, fractures, moldic, and fenestral porosities should be noted. Sandstones normally have porosities which range from about 2% to 36% approximately as follows:

36%	Poorly consolidated, well sorted, fine to coarse—Gulf Coast
30%	Well consolidated, well sorted, fine to coarse—Gulf Coast
25%	Well consolidated, medium sorting—little cement
20%	Well consolidated, medium sorting, well compacted or well cemented—average shallow Oklahoma sandstone
15%	Well consolidated, medium to poor sorting—much cement
10%	Well consolidated, medium to poor sorting—some silica cement—deep burial and well compacted.

5%	Deep burial history—much silica cement—almost metamorphic
2%	Quartzite

Figure 5–7 is Choquette and Pray's (1970) classification of porosities in carbonates, but it is not quantitative. Archie's classification of carbonate porosities consists of two parts, shown in Figure 5–8. Archie maintains that the matrix contains a porosity that is not visible even under 10 power microscope, shown in column A. On top of this porosity is a visible porosity, shown in columns B, C and D. He maintains that the total porosity for the rock can be estimated in the field by simply adding these types of porosities together.

5.4.3 Oil cuts & fluorescence

Positive identification of oil can be difficult based on well logs alone. The definitive test for oil is whether the sample (from cuttings, core, or side

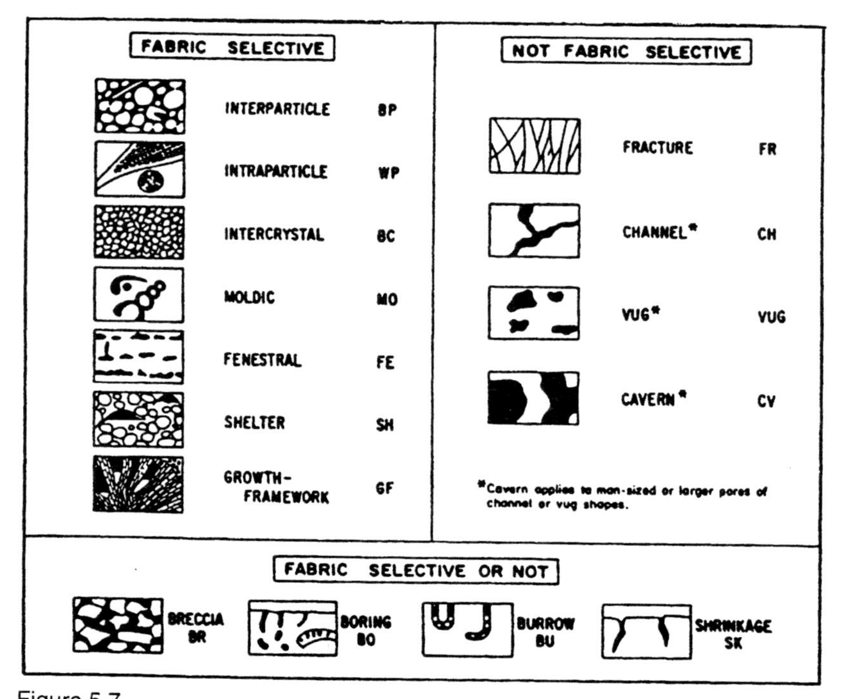

Figure 5.7
Classification of carbonate porosity types. From Choquette and Pray (1970), reprinted by permission of AAPG.

Class	Crystal or Grain Size (Microns)	Usual Appearance (Luster)	Approximate Matrix Porosity % Not Visible (12X-18X)	Visible Porosity (% of Cutting Surface) Size of Pore-mm.			Approximate Total Porosity Percent	
				0.02	0.125–2.0	>2.0		
			A	B	C	D	A+B	A+C
I Compact Crystalline	1000 C 500 M 250 F 125 VF 62 XF 20 SL 4 Li	Resinous	2	e.g.10	e.g.15	†	12	17
		to						
		Vitreous +	5	e.g.10	e.g.15		15	20
III Sucrosic Granular	1000 C 500 M 250 F 125 VF 62 XF 20	Coarsely Sucrosic Granular	5	e.g.10	e.g.15		15	20
		to Extremely Fine	7	e.g.10	e.g.15		17	22
II Chalky	20 SL 4 Li	Chalky +	15	e.g.10	e.g.15		25	30

+Where cuttings are between vitreous and chalky in appearance, designate as I/II or II/I.

†Where pores are greater than about 2.0 mm. and therefore occur at edge of cuttings (e.g., sub-cavernous pores), amount of such porosity is indicated by % of cuttings in an interval showing evidence of large pores.

Figure 5.8
Modified Archie Classification for porosity in carbonate rocks. Modified from Swanson (1981), reprinted by permission of AAPG.

wall sample) fluoresces or gives an oil cut. Live oil (not dead oil) will usually fluoresce a bright yellow gold. If samples of the rock containing oil are ground up into small chips and placed in a cut bottle, and an organic solvent such as trichloroethane or carbon tetrachloride (use in well ventilated areas) is added, the trichloroethane will turn yellow, gold, or brown if oil is present. The oil dissolves in the solvent, and commonly the solvent will then fluoresce. This is one of the best tests for live oil in a reservoir rock. Be aware that most organic solvents are carcinogens and must be disposed of properly. If diesel fuel or Black Magic (to free stuck pipe) have been added to the mud, or if the mud is an oil-base mud, the fluorescence and cut tests are not conclusive, since everything may fluoresce or give a cut. Care should be taken in the use of fluorescence since some carbonate rocks will also fluoresce.

5.4.4 Fractures

Recognizing fractures in a reservoir can be extremely important because they have a strong effect on reservoir performance. Formation fractures may be difficult to distinguish from drilling induced fractures.

Fractured reservoirs may be recognized by:

1. Lost circulation while drilling
2. Cuttings with drusy coating or drusy fracture fills
3. Poor core recovery. Cores will naturally break along native fractures, and core recovery is always poor through fracture zones.
4. Oil-stained surfaces on cores or cuttings
5. Parallel sets of fractures recognized in core
6. Slickensides—polished or smooth rock surfaces caused by the movement of rock against rock on opposite sides of a fault or fracture. They occur naturally on many fault surfaces, but they can also be artificially induced by drilling and coring, especially in shales.
7. Porosity versus permeability plots from cores may show a high degree of scatter.
8. Spikes on well logs such as resistivity, porosity, or dipmeter
9. Fracture identification log. The dual electrodes on the arms of modern dipmeters can be played back for side-by-side comparison. Fracture fills show good correlations that are not parallel to local dip.

10. Acoustic imaging devices, such as the borehole televiewer and electrical imaging may find fractures. Sonic logs occasionally show cycle-skipping across fracture zones.
11. Pressure buildup tests and very rapid pressure response in nearby wells can often identify the presence of fracture systems.
12. Unusual field performance. Very high initial flow and decline rates changing to low flow and decline rates are indicative of fracture depletion changing to matrix depletion through time.

Artificially induced fractures can be identified if:

1. Fractures are conchoidal or very irregular. Natural fractures tend to be planar or in conjugate sets.
2. An uncemented vertical fracture commonly angles in abruptly from the core edge in a downhole direction. Drilling-induced fractures commonly split the core into roughly equal halves, sometimes with a slight rotation about the core axis.

References

Two excellent reference manuals for descriptions of cores and cuttings are:

Sample Examination Manual. Published by the American Association of Petroleum Geologists, P.O. Box 878, Tulsa, Oklahoma, 74101. Originally written for Shell Oil by R. G. Swanson.

AGI Datasheets. Published by the American Geological Institute, 5205 Leesburg Pike, Falls Church, Virginia, 22041.

CHAPTER 6

Pressure and Temperature in the Subsurface

In the subsurface, both temperature and pressure generally increase with depth. Not only do pressures and temperatures themselves increase through conformable vertical sequences, but pressure and temperature *gradients* (Fig. 6–1) commonly increase slightly with depth. Across discontinuities such as unconformities and faults, pressure, temperature, and gradient changes may be abrupt. The overall effect of these increases is not linear; pressure and temperature increase slightly exponentially with depth.

6.1 Hydraulic Pressure in a Wellbore

Figure 6–1 shows an example for a typical Gulf Coast well. The curve at the right shows the pressure gradients that are expected to be encountered in the well. These gradients are expressed either in psi/foot or in mud weight pounds per gallon (ppg). The conversion from psi/foot to ppg is:

1 ppg (pound per gallon) = .0519 psi/ft

Typically, the formation gradient equivalent increases slightly with depth as shown in Figure 6–1. The principal blowout preventer in a well is the hydraulic head caused by the mud in the wellbore. The bottom hole pressure caused by the mud in the wellbore is given by the following equation:

$$P_{wb} = P_s + (TVD \cdot G_m)$$

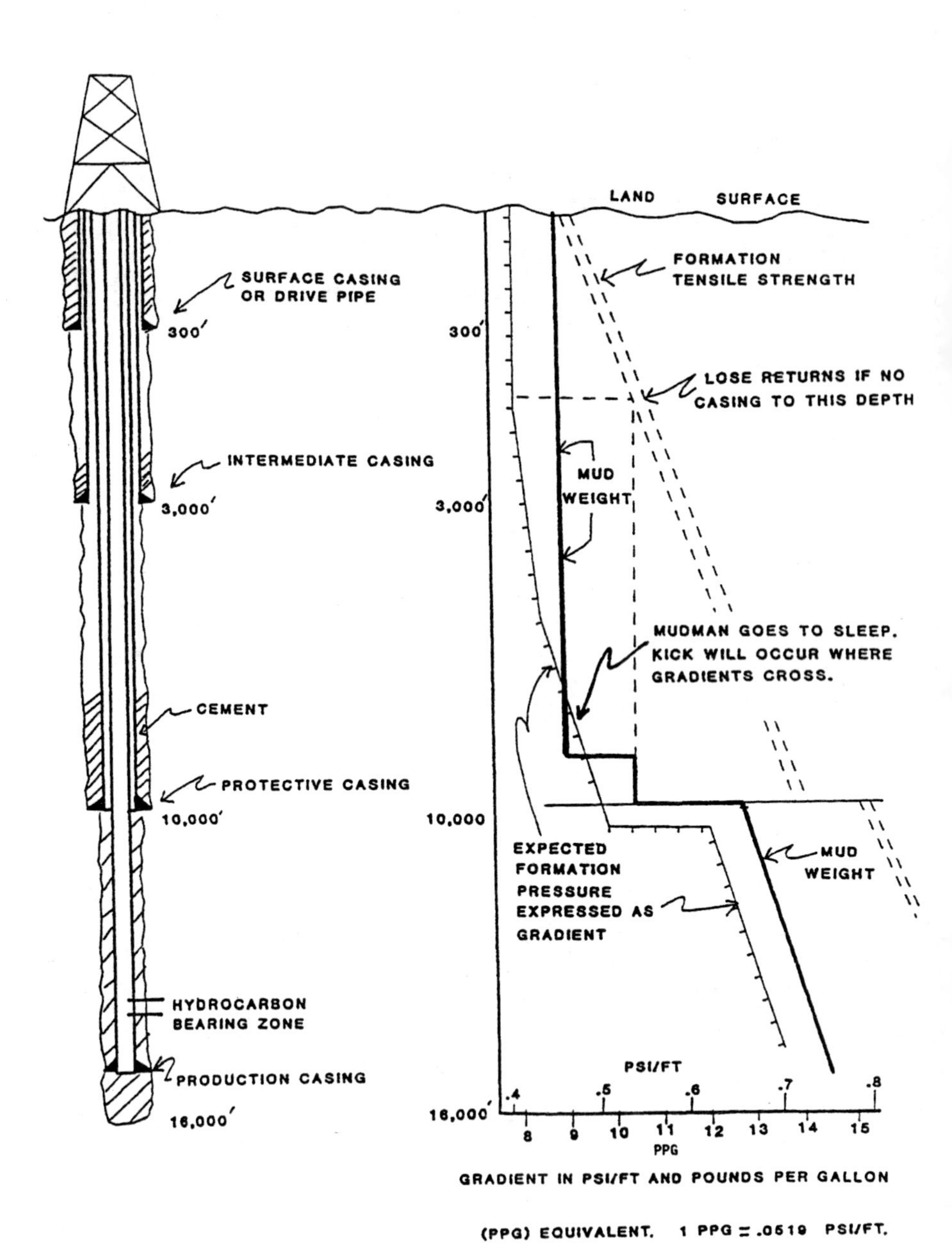

Figure 6.1
Summary of expected mud weight hydraulic equivalents, break down pressures, and casing design for a hypothetical well.

where: P_{wb} = Hydraulic pressure from mud in the wellbore
P_s = Surface (atmospheric) pressure = about 15 psi
TVD = True vertical depth of the well from the drill floor
G_m = Mud weight gradient

Some common gradient equivalents are shown in Table 6–1.

Table 6–1 Common gradient equivalents.

	ppg	Gradient psi/ft
Natural gas (average)	2.1	.11
Diesel fuel (crude oil)	6.7	.35
Fresh water	8.34	.433
Sea water	8.6	.446
Gulf Coast average near surface gradient	9.0	.465
Boundary between "normal" and "overpressures"	9.3	.483
Neat cement	16.8	.87
Lithostatic	22.2	1.15 (2.65 gm/cc)

Example:

Given the following:

12 ppg mud in a straight hole
KB elev. = +300 feet
MD = 9,000 feet

What is the hydraulic pressure caused by the mud at the bottom of the wellbore?

Solution:

$TVD = MD$ in a straight hole = 9,000 feet

$$P_{wb} = P_s + (TVD \cdot G_m)$$

$$P_{wb} = 15 \text{ psi} + (9{,}000 \cdot 12 \text{ ppg} \cdot .0519 \text{ psi/ft/ppg})$$

$$P_{wb} = 5620 \text{ psia}$$

Note: Psia is absolute pressure; Psig is gauge pressure;
Psia = Psig + 15 psi (atmospheric pressure).

If the formation pressure is 5,000 psia, then the drilling fluid overbalances the formation pressure by 620 psi (5,620 – 5,000) and the well should not kick or blowout. In fact, the drilling fluid will move out into the permeable formations and clay minerals will be filtered out at the rock-wall interface to become mud filtrate. Typically, drillers try to keep a 1 ppg overbalance such that the bottom hole mud pressure is 1 ppg (or .0519 psi/ft) greater than the formation pressure.

If the formation pressure at any depth becomes greater than the hydraulic pressure of the mud in the wellbore (crossover point in Fig. 6–1), then formation fluid will flow from the rock formation into the well bore. If incoming fluids contain gas, the gas will expand as it comes to the surface and the bottom hole pressure of the mud will decrease because of gas in the mud. When this occurs, the well is said to be kicking and the well is in danger of blowing out. To prevent a blowout, the mud weight must be increased while the gas is vented at the surface, simultaneously keeping back pressure on the gas in the wellbore. The gas that is vented is commonly a mixture of gas, water, and oil, and because it may be under great pressure, the venting process commonly causes the whole rig to be jarred or kicked.

6.2 Formation Pressure in the Subsurface

Formation pressures in the subsurface must be estimated before a well is drilled so that appropriate mud and casing designs can be formulated for the well. Formation pressures can be estimated by:

$$P_{fm} = P_s + (TVD \cdot G_e)$$

where:

P_{fm} = Formation pressure
P_s = Surface (atmospheric) pressure = about 15 psi
TVD = True vertical depth to the formation
G_e = An estimated (earth) gradient (.465 psi/ft is average Gulf Coast shallow gradient)

6.3 Casing Design and Lost Returns

Casing is one of the most expensive and important parts of any well. Geologists are rarely involved with casing design, yet geologic input is one of the most important criteria for proper casing design. Communication failure between the geologist and engineer on casing design can have disastrous results. Overdesign of the casing can be very expensive.

Underdesign can cause blowouts. The Santa Barbara blowout is a classic example of poor casing design.

In Figure 6–1, pressures and gradients are expected to be considerably higher below an unconformity at 10,000 feet. Below 10,000 feet, mud weights of 13 ppg and higher must be used to keep the well from flowing at the bottom of the hole. However, if the mud weight is simply increased to 13 ppg, without running an intermediate or protective casing string, the formation uphole will actually fracture. That is, the pressure caused by the 13 ppg mud will exceed the tensile strength of the rock, the rock will break down, and mud will flow out into the formation. The well is said to be *losing returns* when this happens. Mud is pumped down the drill pipe, but it does not return to the surface. Instead, it goes out into the fractured formation at relatively shallow depths.

If a kick occurs while losing returns, the rig is in an extremely dangerous position because the driller has no way to maintain fluid pressure on the bottom of the hole except to increase mud weight, which only exacerbates the lost returns problem. If a well is experiencing a kick at one horizon while simultaneously losing returns to another horizon, the well is experiencing a *subsurface blowout*. In the Santa Barbara Channel blowout, the well took a kick while drilling, and as the gas bubble moved to the surface it caused very high pressures on the formation just below the surface casing, which happened to coincide with a fault zone. These high pressures caused a breakdown of the formation, and lost returns occurred along the fault zone. Gas and oil moved up the well from the kicking horizon to the fault plane, and then blew out along the fault plane to the surface. Had the intermediate casing string been set below the fault plane, the well would probably not have blown out.

Thus, the engineer who designs the casing program must have input from the geologist. The drilling engineer typically needs:

1. An estimate of pressure versus depth
2. An estimate of tensile strength or breakdown pressure of the rock
3. Depths of expected unconformities or faults where pressures are likely to change
4. Anything unusual or uncertain that is likely to be encountered in the well. If there is any chance of drilling into shale sheath on the edge of a salt dome, the drilling engineer needs to know. Shale sheath material on the side of salt domes is commonly highly overpressured.

6.4 Temperature Gradients

Temperatures in the subsurface typically increase at rates of about .010 to about .040 degrees F/ft with an average of about .014 degrees F/ft. Locally, temperatures may vary considerably, particularly near active igneous vents or geothermal activity.

A general equation for estimating temperature versus depth is given by:

$$T_{fm} = T_s + (TVD \cdot G_t)$$

where:

T_{fm} = Formation temperature in the subsurface
T_s = Surface temperature which varies by state in the U.S.
T_s = 74°F for Texas and Louisiana
= 60°F in Colorado
TVD = True vertical depth to the formation
G_t = Temperature gradient; G_t ranges from .010°F/ft to .040°F/ft; G_t shallow average = .014°F/ft.

Example:

What is the estimated bottom hole temperature of an average formation at 10,000 feet in Texas?

Solution:

$$T_{fm} = T_s + (TVD \cdot G_t)$$

$$T_{fm} = 74°F + (10{,}000 \cdot .014°F/ft)$$

$$T_{fm} = 214°F$$

Of course, bottom hole temperature can be measured with a down-hole thermometer and or other temperature recording device. Another estimation method (not considered accurate) is to measure the maximum temperature of the mud just before logging. Drilling mud is normally circulated throughout the hole for an extended period just prior to logging to condition the hole and the mud for logging. The circulating mud heats up as it travels in the wellbore, but it never reaches the full temperature of the formation at the bottom of the hole. Thus, the mud temperature recorded on logs is always slightly lower than the true bottom hole temperature.

6.5 Origin of Subsurface Pressures and Overpressures

In normally pressured geologic sections, the formation fluid pressure is a function of geostatic and hydrostatic pressure. *Geostatic pressure* is the pressure derived from overlying rocks. *Hydrostatic pressure* is the pressure derived from fluids in the overlying rocks.

During shallow burial of sand-shale sequences, large amounts of water are expelled as the rocks become compacted. The water escapes to the surface through sandstone aquifers, and the rock remains normally compacted and normally pressured. On deeper burial, shales continue to compact and expel fluids, but as long as aquifers are open to the surface the rock remains normally or near-normally pressured.

Porosity also decreases with increasing depth. Early workers suggested that, because of this trend, there were practical limits to which hydrocarbon reservoirs could exist. That is, a linear extrapolation of these trends suggested that there exists a depth below which porosities should be zero.

"Diagenesis" is considered a bad word among most petroleum geologists because, in most cases, increasing diagenesis is associated with decreasing porosity. For the most part, the statement is true—but not always. A number of workers have shown that in the subsurface there is commonly a depth below which porosity increases somewhat. The zone is referred to as the *zone of secondary porosity* and has been related to two principal causes, clay mineral reactions and organic reactions.

6.5.1 Clay mineral reactions

One of the most fundamental chemical reactions in geology involves the breakdown, or *hydrolysis,* of feldspars to form various clay minerals during chemical weathering. An example is:

$$3KAlSi_3O_8 + 12H_2O + 2H^+ \rightarrow KAl_3Si_3O_{10}(OH)_2 + 6H_4SiO_4 + 2K^+$$

(orthoclase) (muscovite) (illite) (silicic acid)

Further weathering of illite under the right conditions creates kaolinite and ultimately gibbsite, as shown below:

$$2\ KAl_3Si_3O_{10}(OH)_2 + 3\ H_2O + 2\ H^+ \rightarrow 3\ Al_2Si_2O_5(OH)_4 + 2\ K^+$$

(muscovite) (kaolinite)

$$Al_2Si_2O_5(OH)_4 + 5\ H_2O \rightarrow 2\ Al(OH)_3 + 2\ H_4SiO_4$$

(kaolinite) (gibbsite)

The reactions do not necessarily proceed through all of the above steps. For example, orthoclase to kaolinite may bypass illite by the following reaction:

$$2\ KAlSi_3O_8 + 9\ H_2O + 2\ H^+ \rightarrow Al_2Si_2O_5(OH)_4 + 2\ H_4SiO_4 + 2\ K^+$$

(orthoclase) (kaolinite)

Similar chemical equations can be written to show that plagioclase tends to weather to montmorillonite. It has been stated by certain geologists that "one cannot call oneself a geologist if one doesn't know these equations." Although this statement could be debated, the equations are important because:

1. They show what happens to feldspars, the most common mineral group on earth, during weathering.
2. They show the origin of the clay minerals and shale.
3. During intermediate diagenesis, that is, above 80°C, the *equations reverse, or go to the left.*
4. At temperatures ranging from approximately 80°C to 200°C, *enormous amounts of water are generated* by these reactions.
5. *That water must go somewhere.* If that water is confined and cannot escape to the surface, it will migrate into surrounding pore spaces and may create overpressures.

Temperature also increases with depth. Simple thermal expansion of mineral grains and fluids also increases pore pressures in the subsurface.

6.5.2 Organic reactions

Remarkably, 80°C to 200°C coincides approximately with the oil window (see Chapter 8). The chemical reactions that create oil and gas in the subsurface also create organic acids. Both reactions create pressure and fluids much the same way a pressure cooker generates pressure and "good" fluids on heating. The increased pressures dilate the mineral grains and, in many cases, actually fracture the source rock to create avenues of escape for the fluids. Most source rocks (shales and evaporites) have very low permeabilities, and the fracturing process is important as a means of transporting oil and gas out of a source rock.

The process of creating oil and gas in a source rock creates a pressure gradient from the source rock into surrounding rocks. After escaping the source rock, oil and gas migrate to traps through more permeable rocks, such as sandstones and fractured limestones, by moving down this pressure gradient which is normally, but not always, upwards towards the surface. Organic acids dissolve mineral grains along the way and are very important for creating secondary porosity during diagenesis and migration.

These escaping pressures and fluids not only fracture source rocks, they also create excess pore pressures in surrounding rocks. Excess pore pressures tend to dilate fractures and faults and are important for helping to generate migration avenues for oil and gas through rocks that are normally relatively impermeable. If it were not for the fortuitous simultaneous generation of oil, gas, water, organic acids, and pressure, 20th century history might be quite different. Without these simultaneous reactions, it is very unlikely that we would enjoy the huge quantities of inexpensive energy that we have enjoyed during the industrial revolution of this century. The principal mechanism for moving oil and gas from a source rock to a trap would not exist without these simultaneous reactions.

It is not unusual to encounter anomalously high pressures below an unconformity or on crossing a fault. Although any impermeable rock may act as a pressure seal, unconformities seem to be unusually good at it. Rocks below an unconformity have thermal histories that are different from those above, and they are very likely to be overpressured or occasionally underpressured. In setting up an exploration well prognosis, unconformity depths should be identified so that drilling engineers and rig personnel can be prepared for these potential danger zones.

Finally, overpressured reservoirs are likely to behave as depletion drive reservoirs, because an overpressured zone cannot be connected through highly permeable media to the surface. High overpressures are likely to be areally restricted in the subsurface.

6.5.3 Other means of developing overpressures

Overpressures can also be developed by unusually long vertical hydrocarbon columns, particularly by large gas columns. Figure 6–2 shows an extreme example. In this example, the gas-water contact is at –10,000 feet, below a normal (.465 psi/ft) gradient, which means that the pressure at the gas-water contact is approximately 4650 psi. Natural gas has a very low hydrostatic gradient (on the order of .11 psi/ft), depending on the composition of the gas. Assuming the gas has a gradient of .11 psi/ft

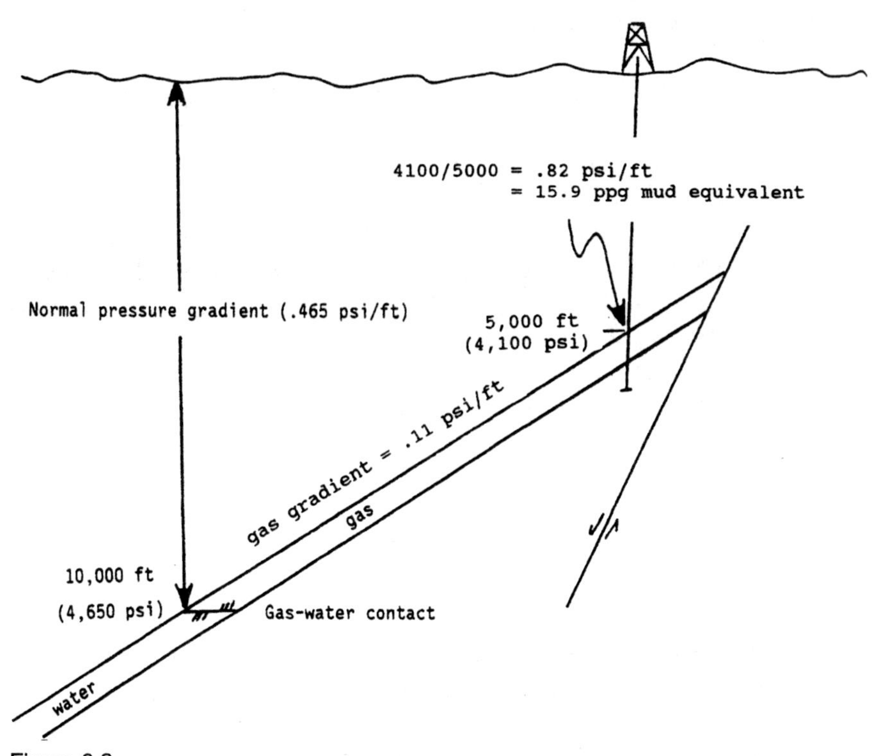

Figure 6.2
Example of overpressures generated at the top of a long gas column in a normally pressured subsurface area.

over the 5,000-foot gas column, the pressure reduction in the reservoir at –5,000 is only 550 psi, meaning that a well that drills into that same reservoir at –5,000 feet will encounter fluid pressures of 4,100 psi. This gradient equivalent is 4100 psi per 5000 feet, or .82 psi/ft, which is a 15.8 ppg mud weight equivalent. In order to maintain a 1 ppg overbalance, this reservoir must be drilled with a 16.9 ppg mud weight.

Long hydrocarbon columns can clearly create very high overpressures and are probably responsible for self-limitation of the size of many reservoirs. That is, as gas continues to build in such a reservoir, the updip limits of the reservoir become more and more overpressured until fractures or faults in the seal become dilated and leak. Hydrocarbons leak through the trap until pressures are diminished, then hydrocarbons start to build again. Most major oil and gas fields around the world are known to leak hydrocarbons at the surface today.

6.6 Underpressures

Unusually low reservoir pressures and low breakdown pressures have been commonly observed in rocks that have been uplifted and dissected by erosion such that fluids have been drained from the formation. Uplift can cause expansion and cooling of overlying rocks, and pressures are commonly reduced through thermal contraction in such rocks. The Navajo Sandstone, for example, is known to have very low breakdown pressure equivalents in parts of the southern Colorado Plateau where deep erosional valleys have drained the sandstone of the majority of its groundwater. Drilling fluids must be aerated (air bubbles pumped into the mud) to decrease its hydraulic gradient to avoid fracturing of the formation and to avoid lost returns.

6.7 Pressure Testing

Many different types of reservoir pressure tests exist. They range from the full-scale drill stem test to the simple repeat formation tester (RFT) commonly run during logging operations. In a full-scale drill stem test, the reservoir is packed off, tubing is run, and reservoir flow rates are measured through several different chokes (restricted flow tubes) to determine open (or maximum) flow potential. More simple tests involve drawdown tests that are commonly run early in the history of a well, or pressure buildup tests that are commonly run after the well has flowed for a period of time. The pressure buildup test is probably the most common pressure analysis test, while the repeat formation tester is the simplest and cheapest.

A short discussion follows on repeat formation testers, pressure buildup tests, and drawdown tests. A discussion of the full drill stem test is left to reservoir engineering.

6.7.1 Repeat formation tester

The repeat formation tester is a simple wireline device. Normally three or more chambers are sealed at surface pressure (15 psi). The chambers and a pressure recording device are lowered into the wellbore. As shown in Figure 6–3, the device measures hydrostatic pressure on the way into the hole. Packers are then set either in open hole conditions or in perforated casing. Each chamber is then opened sequentially and the pressure drawdown plus buildup is recorded. Because the chambers start in a relative vacuum (15 psi), samples of the formation fluid are drawn into the chamber and recovered. In some cases, samples from the RFT have been

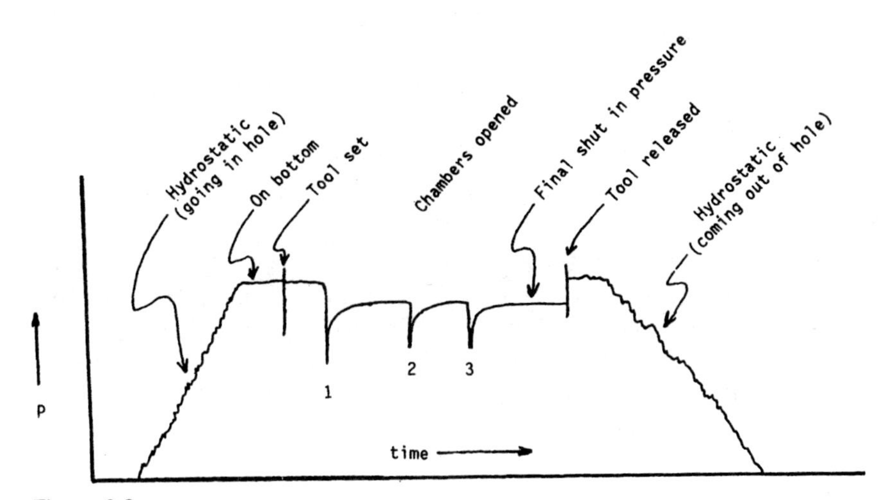

Figure 6.3
Schematic diagram summarizing pressure response for a typical repeat formation test (RFT) with three chambers.

used to confirm production from untested reservoirs and to serve as proof of reservoir producibility to hold a lease.

The principal purpose of the RFT is to determine reservoir pressure and reservoir producibility, and to obtain samples of the formation fluids including oil, gas, and in particular, water for determination of salinity and Rw. The tool also measures final shutin pressure, which is a reasonably good estimate of the formation pressure at the time of measurement. But, because flow periods are very short (on the order of minutes), estimates of permeability and skin effects are crude at best.

6.7.2 Pressure buildup analysis

The most common pressure analysis tests are pressure buildup tests. These tests consist of shutting in a producing well and measuring the pressure buildup in the wellbore as a function of shutin time. Figure 6–4 shows a number of plots of shutin wellbore pressures against log of $((t + \Delta t)/\Delta t)$. This analysis was first introduced to the petroleum industry by D. R. Horner, and the plots are referred to as Horner plots.

Equations for the plots are as follows:

$$P_{ws} = P_i - 162.2\, q\mu B/kh \cdot [\log (t + \Delta t)/\Delta t]$$

where: P_{ws} = wellbore shutin pressure, psia
P_i = initial pressure, psia

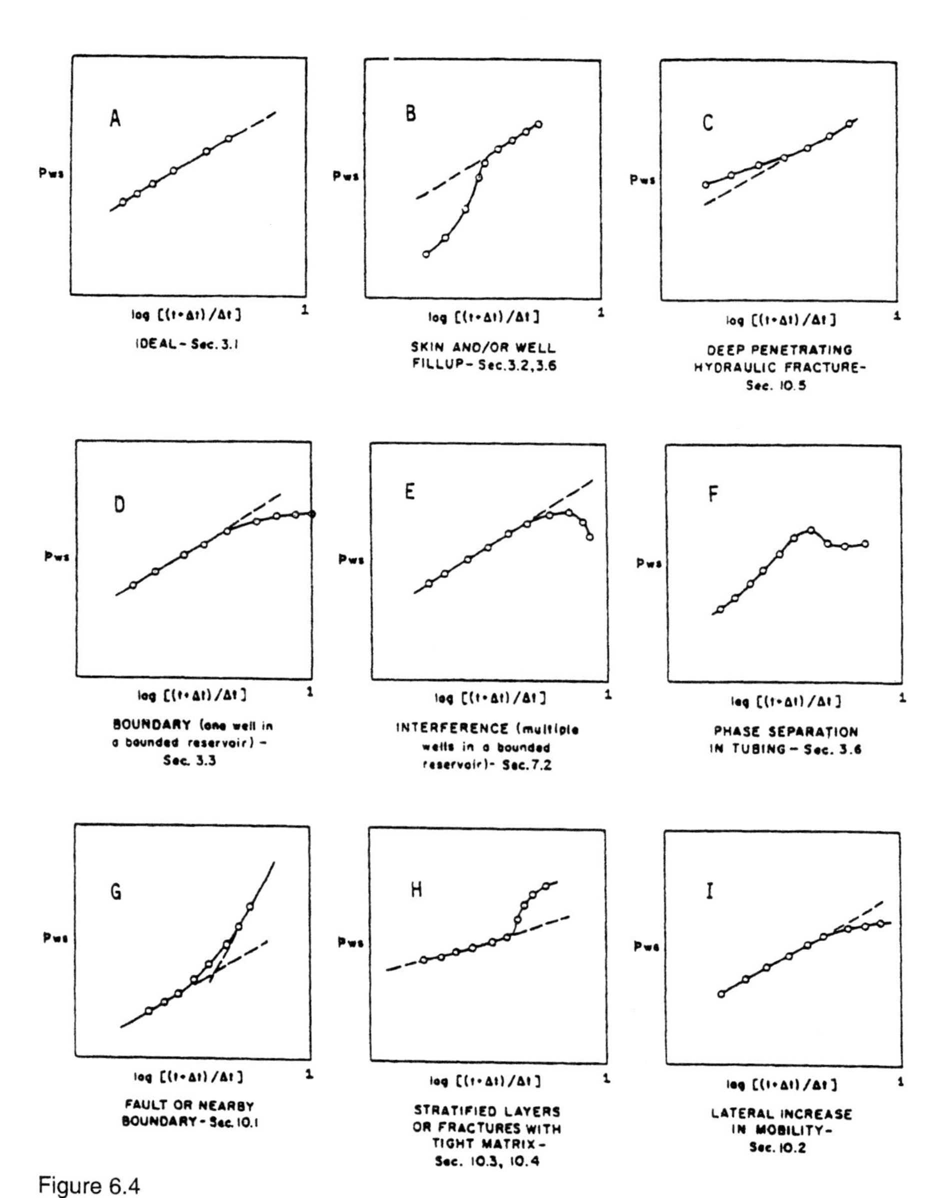

Figure 6.4
Some pressure buildup signatures
From "Pressure Buildup and Flow Tests in Wells," by Matthews, C.S., and Russell, D.G., Monograph 1, Henry L. Doherty Series, 1967, p. 123. Copyright 1967 SPE, reprinted by permission of the Society of Petroleum Engineers.

q = production rate, STB/day
μ = fluid viscosity, cp.
B = fluid formation volume factor, RVB/STB
k = permeability, md.
h = net pay thickness, ft.
t = production time, hrs.
Δt = shutin time, hrs.

This equation is in the form of a straight line:

$$y = mx + b$$

where: $y = P_{ws}$ $\quad m = 162.6\, q\mu B/kh$
$b = P_i$ $\quad x = \log\,[(t + \Delta t)/\Delta t]$

This says that if we plot shutin time as a function of the logarithm of the ratio of t plus Δt divided by Δt, a straight line should occur. This plot may be made on semilog graph paper as shown in Figure 6–4. The slope of this line is defined as:

$$m = 162.6\, q\mu B/kh$$

which may be solved for permeability (really effective permeability) as follows:

$$k = 162.6\, q\mu B/mh$$

As Δt goes to infinity, $[(t + \Delta t)/\Delta t]$ goes to 1. By extrapolation to 1 on the Horner plot, a false pressure (P^*) can be obtained. For an infinite reservoir, $P^* = P_i$, and for a new well in any large reservoir that has seen limited production P^* will be a value close to discovery or virgin reservoir pressure.

Not only is the Horner plot significant for determining effective permeability and initial reservoir pressures, but deviations from a straight line can often be attributed to specific reservoir conditions. Skin effects, effects near the wellbore, usually show up as deviations from a straight line at low Δt values (Fig. 6–4). Reservoir limit tests look for nonlinearity at higher Δt values.

Most of the examples in Figure 6–4 are self-explanatory, but a few are not. It is worth spending a few minutes to think about each of the cases.

Case A This is an ideal example of an *infinite reservoir.* It indicates that, even at large Δt, no reservoir-limiting effects have been identified.

Case B This is the most common occurrence for the early part of most curves and represents a typical example of a strong skin effect and/or a period of time for fill up of the wellbore. The nonlinearity in case B is caused entirely by wellbore or near-wellbore effects. Typically, it means formation damage or some sort of restriction near the wellbore. Formation damage can be caused by mud caking, clay dispersion, cement problems, plugged perforations, paraffin buildup, or any of a number of other geological or mechanical reasons.

A positive skin factor occurs when any sort of flow restriction occurs in or near the wellbore. A negative skin factor implies above-normal reservoir conditions near the wellbore. This can be caused by a number of completion techniques such as acidizing or fracturing.

Case C This situation occurs where flow is better than expected through porous media in radial flow. Typically it means almost open flow conditions and is commonly associated with deep penetrating fractures. It can also occur where the skin effect is negative.

Case D This situation occurs in a restricted reservoir. As the pressure builds up through time, the effective radius increases until it reaches the limits of the reservoir. At that point, the pressure in the wellbore does not continue to build at the same rate, and the curve flattens out to overall reservoir pressure.

Case E Here is a simple case of interference from other producing wells.

Case F This situation occurs where oil flows into the wellbore and the pressure builds up to a maximum. If the pressure is below the bubble point, gas may come out of solution and separate in the tubing, and the resulting bottom hole pressure will show a decrease. The same result will occur if leaks occur in downhole equipment such as packers or tubing.

Case G By conventional graph wisdom, this case has always been puzzling because it appears that the rate of pressure buildup increases with increasing Δt after reaching the fault. In fact, that is not what is happening. It is an artifact of the log $[(t + \Delta t)/\Delta t]$ function. The log $[(t + \Delta t)/\Delta t]$ compresses the right-hand side of the curve, and for large Δt's, the pressure continues to build, but at a slower rate in the presence of the fault than in the presence of an unrestricted reservoir. Case G is the opposite of case I (Fig. 6–4).

It should also be noted that it is possible to calculate the distance to the closest occurrence of the fault effect by finding the Δt that corresponds to the intersection of the two straight line segments of the curve.

Case H This case occurs in a system with two different permeabilities and is similar to case C. The high permeability layer (or fractures) creates the early straight line part of the curve. Then, at higher Δt, the matrix porosity (or tighter part of the reservoir) bleeds slowly into the fractures and is slowly felt at the wellbore.

Case I In this case the permeability of the reservoir increases with distance from the wellbore. Pressure buildup is slow near the wellbore and faster in the more distant, better part of the reservoir.

In general, with increasing Δt, a curve to the right (clockwise) normally means better reservoir conditions with increasing distance from the wellbore. A curve to the left (counter-clockwise) normally means that some sort of restriction is occurring with increasing distance from the wellbore.

6.7.3 Pressure drawdown curves

Similar analyses can be performed on wells that have been shut in for an extended period of time. As the well is opened for production, pressure drawdown can be analyzed in a manner similar, though not identical, to the pressure buildup analysis. Pressure drawdown curves plotted on semilog or linear coordinate paper can be used to estimate permeability, skin effects, and drainage areas. Drawdown analyses are not used as commonly as buildup analyses because the well or group of wells must be shut in for an extended period of time in order for the reservoir to buildup to stable or static reservoir conditions prior to the test. Shutin wells mean lost revenue.

CHAPTER 7

Porosity, Permeability, and Relative Permeability

This chapter is devoted to a discussion of some of the most fundamental concepts in multiple phase flow through porous media. If geologists and engineers are to effectively extract larger volumes of oil and gas from existing reservoirs, it is important that they have an understanding of the pore systems themselves, and how fluids, particularly multiple phase fluids, are distributed and flow through that porous media.

7.1 Some Important Definitions

- **Porosity:** the total sum of openings or voids that occur within a rock. Usually expressed as a decimal or percent, porosity is denoted by the Greek letter phi (ϕ) and is given by the equation:

$$\phi = \frac{\text{bulk volume} - \text{grain volume}}{\text{bulk volume}} \cdot 100$$

Or, in terms of density, porosity may be expressed as:

$$\phi = 1 - \rho_{bd}/\rho_g$$

where: ρ_{bd} = dry bulk density and
ρ_g = grain density.

- **Effective porosity:** porosity that is available for storage of fluids. It does not include isolated (not interconnected) pores.

- **Wettability:** a measurement of the ability of a fluid to coat the rock or mineral surface. In a two-phase system, such as oil and water, the wetting phase forms an acute angle with the rock surface (Fig. 7–1) and the nonwetting phase forms the obtuse angle. Most reservoir rocks are water wet, and most oil source rocks are oil wet. Some carbonate reservoirs have mixed wettabilities.

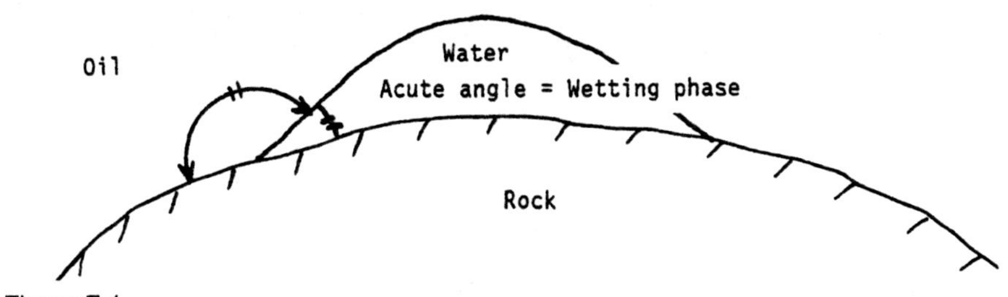

Figure 7.1
Interface between oil and water on a mineral gain in a water wet system. Water is the wetting phase (acute angle) and oil is the nonwetting phase (obtuse angle).

- **Irreducible:** the percentage of effective porosity occupied by wetting phase fluid that cannot be removed on introduction of a nonwetting fluid. In a water wet system, irreducible water is connate water that cannot be driven out, no matter how high the pressure. The only way to remove this water is to alter the wettability characteristics of the system. It is possible to have irreducible oil and residual water in an oil wet system.
- **Residual:** the percentage of nonwetting phase (usually oil or gas) left in the ground after removal of that phase by production or by natural geologic processes.
- **Recovery efficiency (RE):** is equal to one minus the residual (as a decimal). Both recovery efficiency and residual percentages may be changed by increasing pressure, water flooding, or a number of secondary and tertiary recovery techniques.
- **Permeability (k):** a measure of the ease with which a fluid can flow through a material. It is defined by the equation shown in Figure 7–2 and is expressed in darcies or millidarcies. Permeability is commonly measured in the laboratory by forcing helium or mercury through a core inserted into a cylinder similar to that shown in Figure 7–2.

FLUID FLOW RATE - "DARCY'S" LAW

$$q = k \cdot \frac{1}{\mu} \cdot \frac{\Delta P}{L} \cdot A$$

q = fluid flow rate (cm^3/sec)

k = permeability (darcies)

μ = fluid viscosity (centipoises)

$\Delta P/L$ = pressure drop per unit of distance in the direction of fluid flow (atmos/cm)

A = cross sectional area of rock normal to the direction of fluid flow (cm^2)

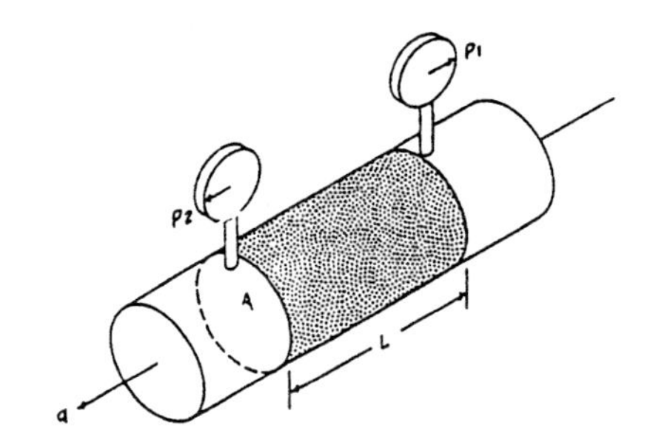

Figure 7.2
Summary of Darcy's Law for linear flow.

- **Permeability for radial flow:** The diagram in Figure 7–3 is a rearrangement of Darcy's Law for radial flow over an effective (*e*) distance from a wellbore (*w*) and the associated pressure drop for that effective distance. Note *kh* (permeability times thickness). *Kh* is the most important indicator of how well an individual well will produce, and *kh* isopachs can be made to outline areas of high and low production capacities.

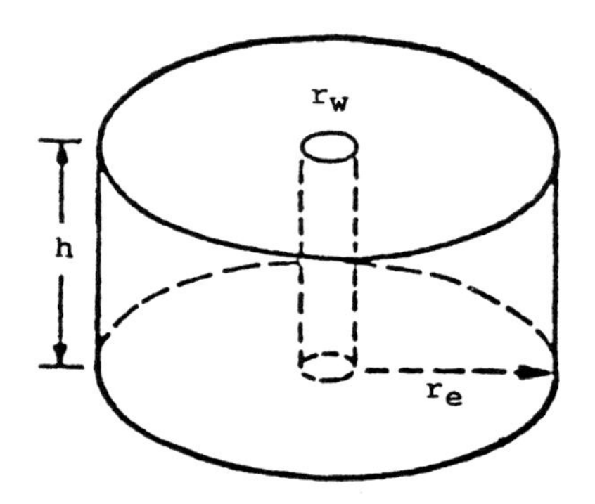

h = thickness

r_e = effective radius

r_w = inner radius

p_w = pressure at the wellbore

p_e = pressure at r_w

$$q = \frac{2 \cdot \pi \cdot k \cdot h}{\mu} \cdot \frac{(p_e - p_w)}{\ln(r_e/r_w)}$$

Figure 7.3
Summary of Darcy's Law for radical flow.

7.1.1 Relative permeability (k_r)

In a rock that contains more than one fluid, relative permeability is the percentage (or decimal) of the total permeability of the rock to that fluid (see Fig. 7–4).

7.1.2 Effective permeability (k_e)

In a rock that contains more than one fluid, the permeability of the rock to any one of the fluids is known as the effective permeability for that fluid and is given by:

$$k_e = k \cdot k_r$$

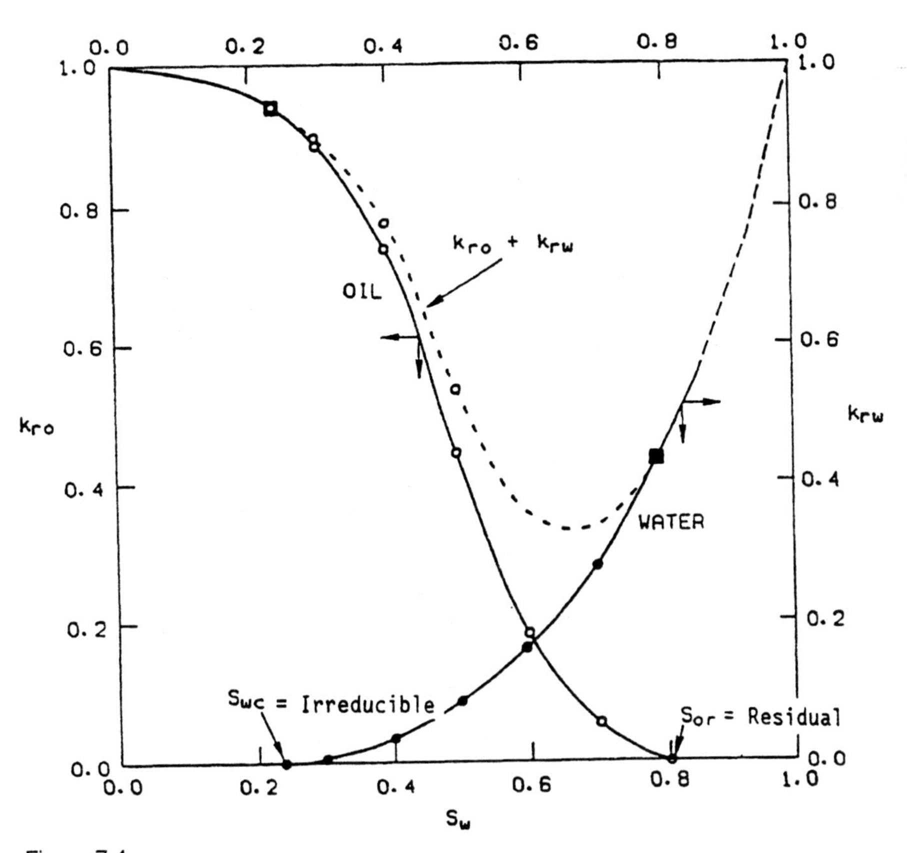

Figure 7.4
Relative permeability of a water wet, oil-water system. Note that water is irreducible, oil is residual. For an oil wet system, oil is irreducible and water is residual.

7.2 Discussion

A number of extremely important concepts are derived from these relationships. The relative permeability diagram (Fig. 7–4) shows that when two or more fluid phases are present, the phases interfere with each other. At an S_w of 0.65 the sum of the effective permeabilities is less than 40 percent of the total permeability of the rock. The diagram also shows how oil and water flow into a wellbore relative to each other as a function of water saturation (S_w), the same water saturation that was measured using the Archie equations in Chapter 4. Thus, water saturation is important not only for reserve calculations, but for prediction of reservoir performance. The diagram shown is for a water wet, oil-water system. A relative permeability diagram for a gas-water system would be similar.

Figure 7–5 shows diagrammatically what happens in the pores for these different water saturations. For very low water saturations, water (the wetting phase) coats the grains in layers that are several molecules thick. Oil flows to the wellbore in continuous phase, and oil flow is affected only slightly by the small amount of water coating the grains.

At intermediate water saturations, oil will exist in two forms, in continuous phase and in discrete or isolated, noncontinuous globs. The discrete globs will become trapped behind small pore throats and will not flow to the wellbore. Continuous-phase oil will continue to flow to the wellbore but will become more restricted as water saturation increases. At very high water saturations, the discrete globs of oil are simply caught behind small pore throats. For a given driving pressure, the oil simply cannot "pop" through the pore throats. To help the oil globules pop through the pore throats, the following can be done:

1. Increase the driving pressure by water injection or gas injection from a nearby well.
2. Reduce the surface tension of the oil. This can be done by adding surfactants, or by heating the oil by steam or fireflood.
3. Dissolve the oil in water, steam, CO_2, or an organic solvent.
4. Expand the pore throats by dissolving the rock (acidizing).
5. Change the wettability characteristics of the rock. This is very difficult (perhaps impossible) to do, and it might make things worse instead of better. Oil would become irreducible and it might become impossible to remove from the rock.

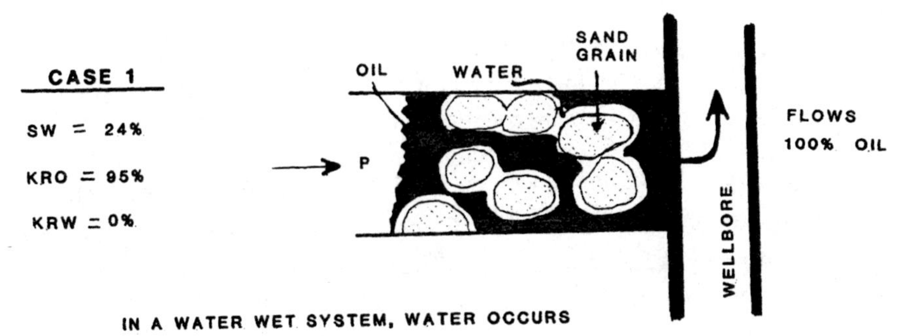

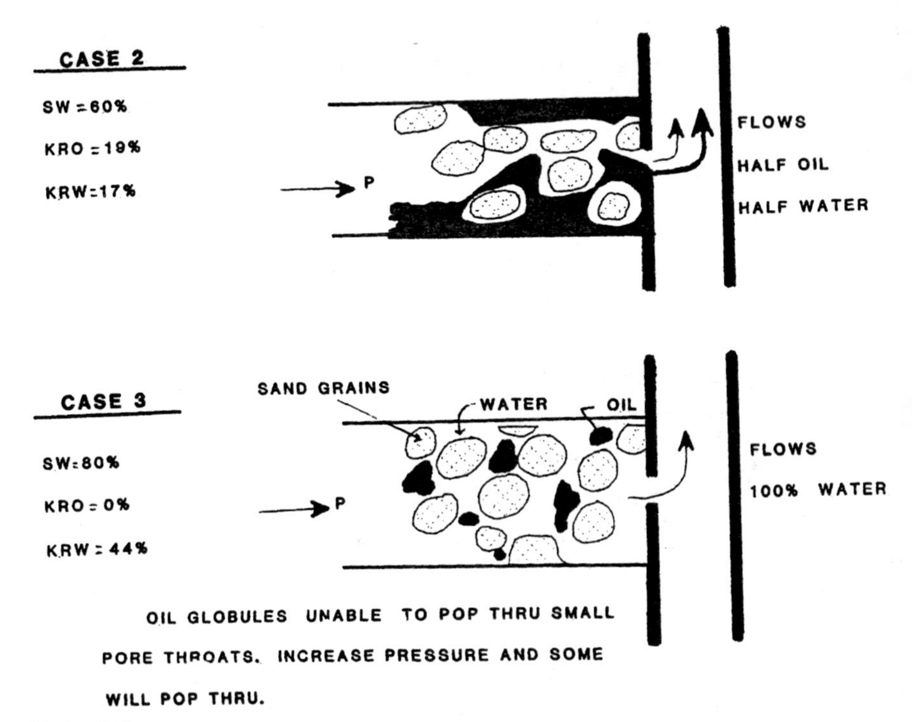

Figure 7.5
Three cases showing relative permeability as a function of S_w in a water wet system.

All of these methods are designed to reduce the residual amounts of oil and gas that are left in the ground. These relative permeability concepts are the fundamental concepts behind all secondary and tertiary recovery techniques (see Chapter 15).

7.3 Migration from Source Rock to Reservoir Rock and Trap

7.3.1 Primary and secondary migration

Relative permeabilities are also important in migration of oil and gas. *Primary migration* is considered by organic geochemists to be migration from a relatively impermeable source rock (such as shale or evaporite) to the primary migration paths such as faults, fractures, or permeable carrier rocks such as sandstones. Many people today feel that most source rocks are oil-wet and that primary migration occurs through microfractures that open and close as extreme pressures are created by the generation of oil and gas from kerogen in the source rock (see Chapter 8).

Secondary migration is migration to a trap that occurs along paths of established permeability such as faults, fractures, and permeable carrier rocks. *Remigration* refers to oil or gas that are remobilized when a trap is sprung.

7.3.2 Capillary pressures and migration by density differential

An important problem occurs when one considers how oil migrates into a water wet reservoir rock. If small globules of oil will not migrate out of a reservoir rock, given the driving forces that occur near a wellbore, why should they move upward through a reservoir rock to a trap, given the very small density differential between oil and water?

One can observe how oil enters a water wet reservoir rock by looking at capillary pressure curves or mercury injection pressure curves. Figure 7–6 shows a mercury injection curve for an excellent reservoir rock (g), an average reservoir rock (i), and a poor reservoir rock (p). The curves show the percentage of rock saturated with oil (or mercury) at different injection pressures.

In excellent reservoir rocks, such as rock g in Figure 7–6, there is little resistance to oil or gas movement. In such a rock, oil or gas will flow easily through very large pore throats. However, to have such an avenue all the way from source rock to trap is rare.

Intermediate reservoir rocks, such as rock i in Figure 7–6, are common. Notice the knee on the curve. The knee represents a threshold pressure below which oil has trouble entering the rock. Small amounts of oil

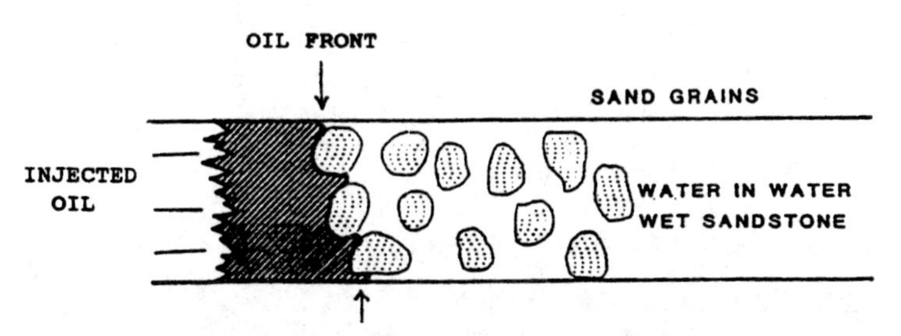

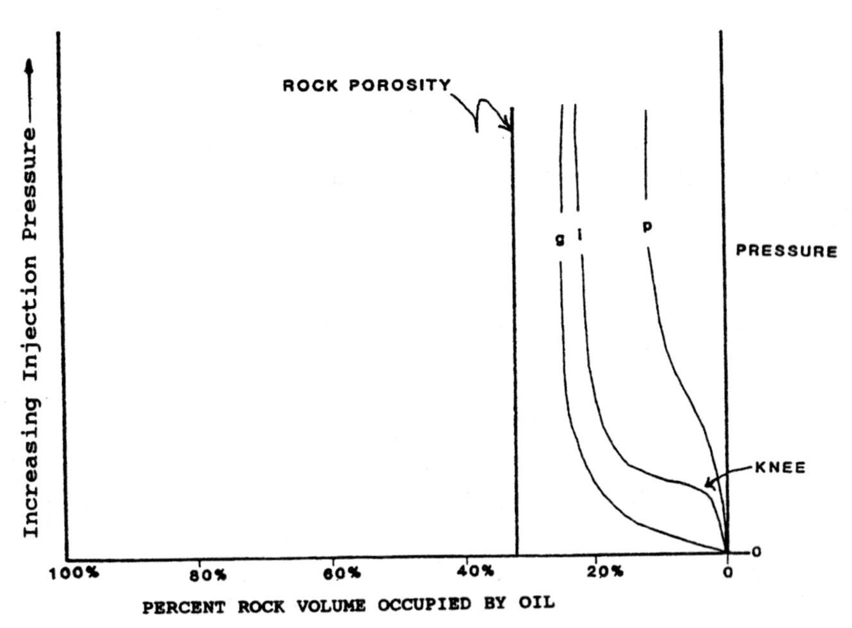

Figure 7.6
Injection pressure versus percentage of the rock volume occupied by oil for a good (g), intermediate (i), and poor (p) reservoir rock. Pore throat size distributions can be derived from these profiles.

or mercury may enter through very large pore throats, but large pore throats are rare. At the knee pressure, the oil starts to pop through the pore throats. In a fine-grained, well-sorted rock, the size of the pore throats may be quite uniform, and it is at this threshold pressure that the oil front begins moving forward. Once through, the oil is in continuous phase and will continue to move, as long as the oil front continues to move.

In the case of the poor reservoir rock (p), it is difficult to inject oil or mercury into the rock, and there is no threshold pressure after which large quantities of oil may enter the rock.

Back to the original question, will oil move by density differential alone? The answer is yes for open fractures or a truly excellent reservoir rock such as g, but to have such an avenue from source rock to trap is almost impossible. Along the way, oil and gas must move through i and p type rocks.

For i and p type rocks, small globules of oil will not move by density differential alone. Something has to push the oil. That oil must be pushed by one or more of these features:

1. Strong pressure differential from source rock to trap
2. Strong water flow from source rock to trap
3. A long, continuous-phase hydrocarbon column.

When enough oil accumulates in continuous phase over a large vertical column, buoyant forces may be transmitted through that column such that pressure at the upper end of the column (see Chapter 6) may be great enough for the leading edge of the oil to pop through the pore throats. For a well-sorted, medium-grained sandstone composed of quartz spheres in tetrahedral packing, a continuous-phase oil column of approximately 25 feet is required before the oil column will migrate by density differential alone.

As oil accumulates in a trap, it must displace water downward at the oil-water contact by density differential alone. Because the density differential between oil and water is small, and because the hydrocarbon column is small near an oil-water contact, oil has trouble displacing water near the oil-water contact. It is particularly unable to displace water through small pore throats because of capillary pressure forces. Even if it could, the interfacial tension is such that water tends to be pulled upward into the oil zone through small pore apertures by capillary pressure.

The important point is that the oil-water contact is always a gradational contact. And the smaller the pore throats, the larger the transition zone will be. Siltstone transition zones are commonly greater than 50 vertical feet thick, and even where the rock is fully saturated with oil, irreducible water saturation may be very high.

7.3.3 Porosity in very fine-grained sedimentary rocks

Siltstones, claystones, and shales commonly contain large quantities of water that are loosely attached to clay minerals. Clay minerals are composed of sheet structures with large electrostatic charges.

Water is a highly polar molecule because hydrogen atoms tend to rotate in planes not 180 degrees from each other, but 120 degrees from each other (Fig. 7–7). Thus, single water molecules have a positive side and a negative side, and as shown in Figure 7–8, they attach themselves in varying degrees to the charges on the clay minerals as follows:

- *Attached:* Chemically incorporated into the lattice of the clay mineral, such as montmorillonite
- *Strongly attached:* Strongly adsorbed water layers on the edges of the clay mineral
- *Moderately attached:* Highly structured water that is moderately adsorbed on the grain
- *Loosely attached:* Moderately structured water that is loosely adsorbed on the grain
- *Unattached:* Unstructured or free water in the middle of the pore spaces

Unconsolidated mud is reported to have porosity in the range of 80 percent. Consolidated shales are reported to have very high porosities (in

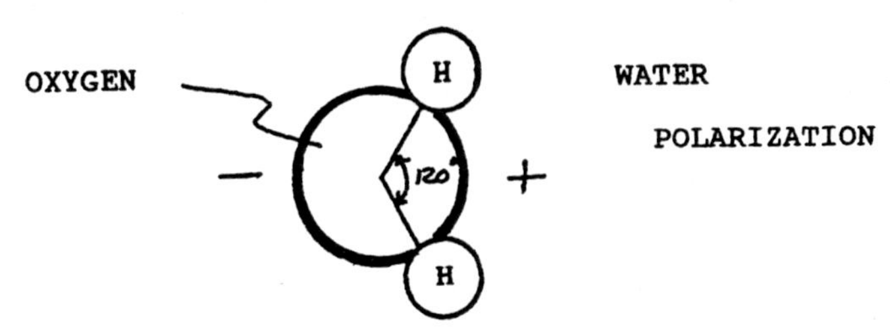

Figure 7.7
Polarity of water molecule caused because positive hydrogen atoms do not rotate at 180 degrees from each other.

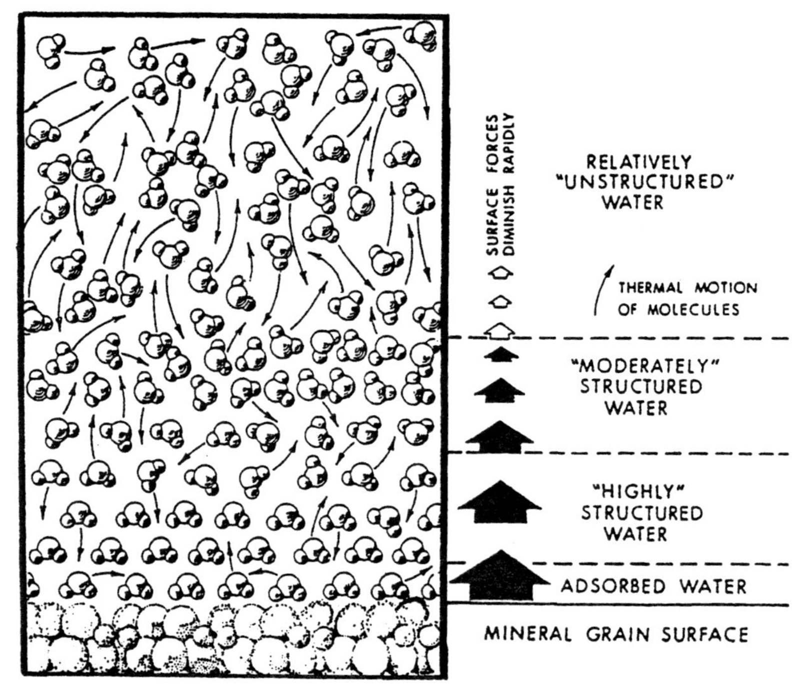

Figure 7.8
Dynamic structuring of water in close proximity to a mineral grain due to adsorption and polarity of water molecules. From Hinch (1980), reprinted by permission of AAPG.

the range of 10 to 40 percent), and porosity logs show shales to have relatively high porosity. It is difficult to define porosity in very fine-grained rocks because of the gradational nature of the clay-water-pore interface.

CHAPTER 8

Oil and Gas Chemistry

This chapter begins with some definitions that are fundamental to the petroleum industry, and concludes with a discussion of the generally accepted modern origin of hydrocarbons; destruction of hydrocarbons, either partial or total; phase relationships in oil and gas reservoirs; and formation volume factors for oil and gas.

8.1 Some Important Definitions

- **Petroleum:** From Latin *petro* (rock) + *oleum* (oil); gaseous, liquid, or solid mixture of many hydrocarbons and hydrocarbon compounds occurring naturally in rocks.
- **Hydrocarbon:** Any compound of hydrogen and carbon; can be gas, liquid, or solid. Does not generally include fats, amino acids, carbohydrates, alcohols, and other organic molecules.
- **Natural gas:** A petroleum that is a gaseous mixture under "normal" surface conditions of pressure and temperature.
- **Crude oil:** A petroleum that is a *liquid mixture* under "normal" surface pressure and temperature conditions.
- **Standard conditions:** Strict chemical definition is zero degrees C and 760 mm. mercury.
- **"Normal" surface conditions:** Average surface conditions; these vary from state to state. Approximately:

  Atmospheric pressure = 14.7 to 15.025 psia = about 15 psi

Normal temperature = 60 to 74 degrees F.

- **Condensate:** Liquid hydrocarbons precipitated from natural gas during the change from reservoir temperature and pressure to normal surface conditions.
- **Solution gas:** Dissolved natural gas that comes out of solution during the change from reservoir conditions of temperature and pressure to normal surface conditions.
- **Vitrinite:** One of the coal macerals—dominant organic constituents of most humic coals.
- **Maceral:** Optically identifiable component of coal. Minerals are to rocks as macerals are to coal.
- **Kerogen:** Three definitions:

  1. Originally: The organic matter in oil shales that yielded oil on heating.
  2. Strict chemical definition: Disseminated organic matter in sedimentary rocks that is insoluble in nonoxidizing acids, bases, and organic solvents.
  3. Most common definition: Reaction products and intermediaries of diagenesis that are no longer organic but are not yet petroleum. It is the precursor to almost all oil and gas.

- **Thermal degradation:** Same as cracking. Breaking down of long-chain hydrocarbons to shorter-chain hydrocarbons with liberation of free (commonly fixed) carbon. Ultimate cracking gives all methane and fixed carbon. *Example:* pentane to methane plus carbon.

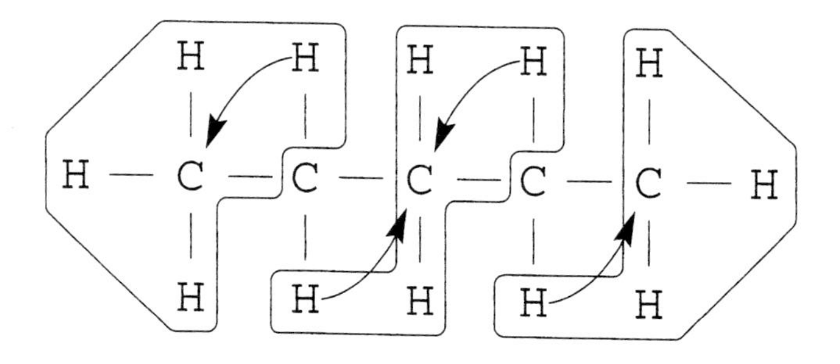

$$C_5H_{12} \longrightarrow 3\,CH_4 + 2C$$

- **Oxidation:** Burning of hydrocarbons in the presence of oxygen resulting in generation of energy, carbon dioxide, and water. In your automobile, this is combustion. For example, octane plus oxygen goes to carbon dioxide and water.

$$2\ C_8H_{18} + 25\ O_2 \longrightarrow 16\ CO_2 + 18\ H_2O$$

- **NSOs:** Nitrogen, sulfur, and oxygen impurities attached to hydrocarbon chains. Along with incomplete combustion, they result in emission of impurities such as NO_2, SO_2, CO, and others into the atmosphere.
- **API gravity:**

$$\text{API gravity} = \frac{141.5}{\gamma_0} - 131.5$$

where: γ_0 = oil specific gravity

> 50 = Distillate
37 = Favored
> 30 = Light
22–30 = Medium
< 22 = Heavy
< 10 = Extra heavy
< 6 = Not producible

8.2 Classification of Hydrocarbons

Hydrocarbons are typically classified according to the following scheme:

I. **Aliphatics:** Chains, branched chains, and cyclic non-aromatics. H/C ratio is approximately 2/1.

A. **Alkanes:** Chains and branched chains—single carbon bonds, very stable.
Composition = C_nH_{2n+2}
See Figure 8–1 and Table 8–1 for boiling and melting points.

B. **Alkenes:** Chains and branched chains with at least one double bond present.
Composition = C_nH_{2n}
Double bond is unstable in the subsurface. See Figure 8–2.

C. **Alkynes:** Chains and branched chains with at least one triple bond present.
Composition = C_nH_{2n-2}
Triple bond is very unstable in the subsurface. *Example* = acetylene.

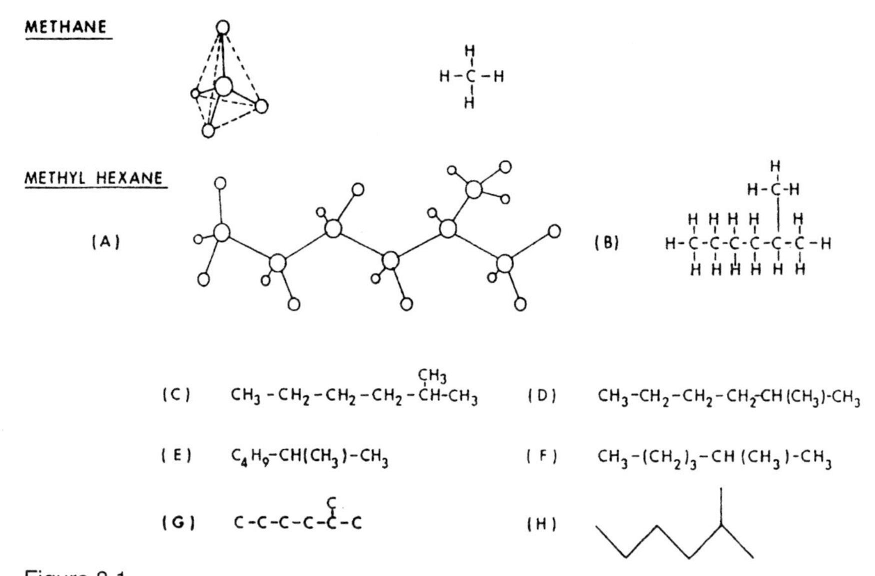

Figure 8.1
Some conventions for representing typical *n*-alkane organic molecules. (A) through (H) are all representations of 2-methyl hexane. From Barker (1982), reprinted by permission of AAPG.

D. **Cyclic aliphatics:** Ring structures not including benzene ring. See Figure 8–2.

II. **Aromatics:** Must have a benzene ring attached somewhere. H/C ratio = ≈ 1, though commonly attached to aliphatics with higher H/C ratio. See Figure 8–2.

Gasoline is normally composed primarily of aliphatics in the range of C_8 to C_{16}. Kerosene is in the range C_{12} to C_{18}, lubricating oils are C_{18} and higher, and paraffins are in the C_{20} to C_{34} range.

As you can see, by simply rearranging side chains on the alkanes, the number of possible combinations of hydrogen and carbon with the same chemical composition are almost limitless. As an example, $C_{30}H_{62}$ has over four billion different arrangements of hydrogen-carbon combinations.

In the subsurface, alkanes, sheets of aromatics, and some cyclic aliphatics are quite stable. Unsaturated hydrocarbons, those with double and triple bonds (alkenes and alkynes), are not stable and are not common in the subsurface. Most organic chemistry courses introduce hydrocarbon nomenclature and stress the incredible diversity of the chains, rings, and sheet structures. They tend to focus not on the hydrocarbons, but rather on the chemistry of the attached functional groups because the attached functional groups are the least stable and most chemically active part of most organic systems. Although nitrogen, sulfur, and oxygen (NSOs) are considered impurities in hydrocarbons, they are the foundation for the formation of proteins, amino acids, lipids, fats, and various chains that are the basis of life. Some of the more common functional groups are summarized in Table 8–2.

Table 8–1 Common names and physical constants for the normal alkanes.

Name	Formula	Melting Point (°C)	Boiling Point (°C)	State at 1 Atmos. & 25°C
Methane	CH_4	–183	–162	Gas
Ethane	C_2H_6	–172	– 89	Gas
Propane	C_3H_8	–187	– 42	Gas
Butane	C_4H_{10}	–135	– 0.5	Gas
Pentane	C_5H_{12}	–130	36	Liquid
Hexane	C_6H_{14}	– 94	69	Liquid
Heptane	C_7H_{16}	– 91	98	Liquid
Octane	C_8H_{18}	– 57	126	Liquid
Nonane	C_9H_{20}	– 54	151	Liquid
Decane	$C_{10}H_{22}$	– 30	174	Liquid
Undecane	$C_{11}H_{24}$	– 26	196	Liquid
Dodecane	$C_{12}H_{26}$	– 10	216	Liquid
Tridecane	$C_{13}H_{28}$	– 6	234	Liquid
Tetradecane	$C_{14}H_{30}$	6	251	Liquid
Pentadecane	$C_{15}H_{32}$	10	268	Liquid
Hexadecane	$C_{16}H_{34}$	18	280	Liquid
Heptadecane	$C_{17}H_{36}$	22	303	Liquid
Octadecane	$C_{18}H_{38}$	28	303	Solid
Nonadecane	$C_{19}H_{40}$	32	330	Solid
Eicosane	$C_{20}H_{42}$	36	—	Solid

From Barker (1982), reprinted by permission of AAPG.

ALKENE (C_nH_{2n})

$H-C\equiv C-H$

ALKYNE (C_nH_{2n-2})

CYCLOBUTENE

CYCLOPENTENE

CYCLIC ALIPHATICS (C_nH_{2n})

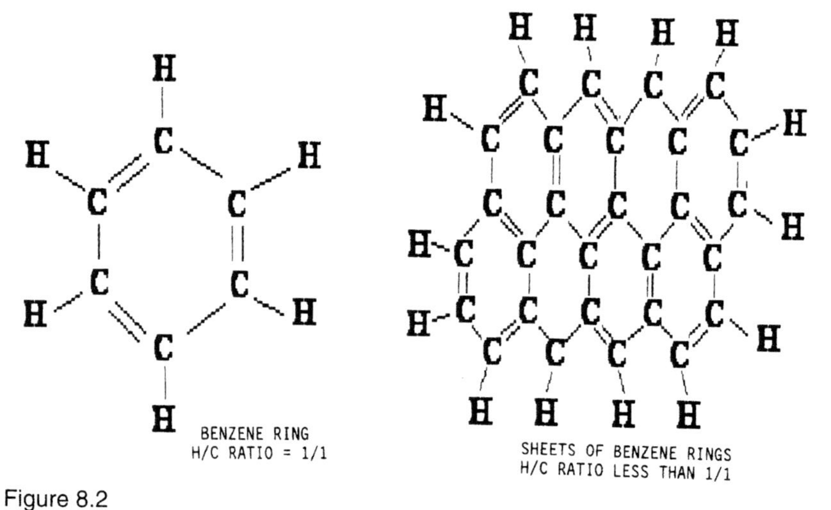

Figure 8.2
Graphic representation of some typical alkenes, alkynes, cyclic aliphatics, and benzene rings.

8.3 Origin of Hydrocarbons

Figure 8–3 is a summary diagram showing a number of processes that occur in the transformation of organic material to oil. The processes are complicated, and not well understood. The first process that occurs as organic material dies and accumulates on the surface of the land or in the ocean is biochemical degredation or just plain rotting. Unstable

Table 8–2 Some common functional groups.

Chemical Symbol	Name
	Hydrocarbon Groups
$-CH_3$, $-C_2H_5$, $-C_3H_7$	Methyl, ethyl, propyl
$-C_6H_5$	Phenyl
–R, –R' –R"	Any alkyl group or predominantly hydrocarbon group
–Ar	Any predominantly aromatic group
	Oxygen-Bearing Groups
–OH	Hydroxyl ("alcoholic" if linked to an aliphatic group; "phenolic" if linked to an aromatic group)
–C(=O)OH	Acid
–C(=O)OR	Ester
C = O	Carbonyl
R–C(=O)–R'	Aldehyde if R' is H; ketone if R' is an alkyl group
R–O–R'	Ether
$-OCH_3$	Methoxyl
	Nitrogen-Bearing Groups
$-NH_2$	Amino
–C=N	Nitrilo
	Sulfur-Bearing Groups
–SH	Mercaptan (thiol)
R–S–R'	Sulfide (thioether)

From Barker (1982), reprinted by permission of AAPG.

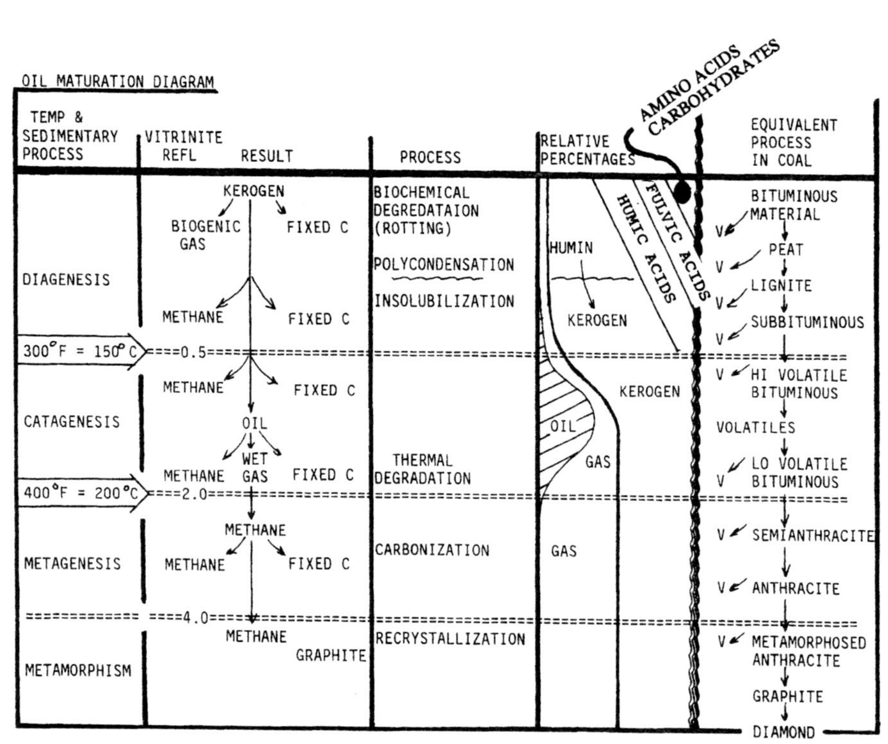

Figure 8.3
Organic maturation diagram showing principal processes for the origins of kerogen, oil, gas, and coal. Please note the heavy, wavy vertical line. This is meant to imply that coal rank is not perfectly analogous to kerogen maturation, but the maturation processes are similar. Modified from Tissot & Welte (1978), reprinted by permission of Springer-Verlag.

chemical constituents break down and/or start to dissolve in water. Carbon dioxide, methane, and smelly sulfur gasses are given off. This is the origin of biogenic gas that occurs in some very shallow reservoirs. It will show up as a bright spot on seismic lines even in very low concentrations.

As the organic material decomposes, NSO functional groups are expelled from the hydrocarbon chains, and polycondensation occurs as the functional groups are eliminated. That is, hydrocarbon chains link together to form longer chains as attached functional groups are expelled. This transformation leads to the development of complex organic acids such as fulvic acids, humic acids, and what soil scientists call humin.

The next process is insolubilization, and this seems to be quite an important process. The transformation from humin to kerogen is a process whereby anything that is in any way soluble in the subsurface gets dissolved and eliminated. Thus, kerogen is essentially an insoluble residue. In fact, kerogen is concentrated for laboratory studies by dissolving the country rock with hydrochloric and hydrofluoric acid. Kerogen seems to be the precursor to virtually all oil, and most natural gas.

Small amounts of gas are clearly generated during the early stages of diagenesis (biogenic gas), before kerogen is generated. Although small amounts of methane are known to outgas from the earth's mantle, the vast majority of scientists today play down the inorganic origin of hydrocarbons, especially long chain hydrocarbons. The amount of methane that has outgassed from the mantle is probably great considering all of geologic time, but is insignificant in terms of today's recoverable gas.

8.3.1 Pressure cooker analogy

The generation of oil and gas from kerogen has been likened to the creation of a good stew from vegetables and a roast in a pressure cooker. Until the heat is turned on, the vegetables and meat just sit in the pressure cooker and don't do very much. Let them sit for a couple of days, and they will give off a certain amount of gas, but not much oil. Turn on the heat, and things begin to cook. Most scientists today feel that oil is generated at rates approximated by the Arrhenius Equation:

$$K = A\, e^{-Ea/RT}$$

where: K = rate
A = constant
Ea = activation constant
R = gas constant
T = temperature

The equation is important because it shows that the rate of oil generation is exponentially dependent on temperature and linearly dependent upon time. Most scientists today feel that at low temperatures (say 80–100°C), little oil is generated unless large amounts of geologic time are involved. But, turn the heat up and things start to cook quickly. The pressure cooker analogy is good because as heat is increased, the vegetables and roast start to cook. They give off nice smelling vapors that would not escape if the pressure cooker did not have a bobber, which is a pressure release mechanism at the top. And of course, if the pressure is not released by the bobber or a release gasket, the pressure cooker will explode. Kerogen in the subsurface behaves in much the same manner. It does not explode, but as pressure is increased, the tensile strength of the surrounding rock may be exceeded, and oil and gas will fracture the surrounding rocks and force themselves out of the source rock. It is this primary migration mechanism that creates microfractures throughout many source rocks and allows the gas and oil to escape through an otherwise impermeable source rock.

Thus, at temperatures of 80–100°C, small amounts of oil can be generated. At temperatures of 150°C, oil will normally be generated very actively if the source kerogen is of the correct type. At temperatures near about 175°C, the long-chain hydrocarbons, namely oil, start to thermally degrade or crack. That is, the long chains simply break into shorter-chain hydrocarbons with the liberation of carbon that eventually becomes fixed. Ultimately, at temperatures near 200°C, all of the long-chain hydrocarbons will completely crack to a point where the only remaining constituents are pure methane and fixed carbon.

The temperature range between approximately 100° to 200°C is called the *oil window*. Source rocks that have been heated to less than 100°C are considered to be immature, source rocks that are in or have been in the range of 100° to 200°C are considered to be mature, and source rocks that have been heated to greater than 200°C are considered to be overmature. In the pressure cooker, if you turn the heat up too high, or cook the vegetables and meat too long, they turn to black messes of carbon and volatiles are driven off. There is definitely an optimum range of time and temperature to which the roast or kerogen should be cooked. Care should be taken to note that the 100°C lower boundary is approximate. Certain source rocks seem able to generate large amounts of oil at temperatures below 100°C if they are given enough time. Other kerogens will not begin to produce oil until heated well past 100°C.

A number of different maturation indices exist, including vitrinite reflectance and conodont color, and many oil companies have their own methods for making these estimations.

Some important conclusions from this discussion are:

1. Oil and gas reservoirs that occur in the upper part of the oil window (say near 175°C) are likely to be very tricky. Oil at this temperature will be partially thermally degraded (cracked) and the reservoir is likely to be either:
   (a) An oil reservoir with an extremely high gas-oil ratio, or
   (b) A gas reservoir with an extremely high condensate yield.

In either case, determination of formation volume factors can be very tricky, and reservoir performance can be difficult to predict.

2. At temperatures of over approximately 200°C, oil should not be expected. The next oil crisis will not be solved by drilling deeper. Oil simply does not exist in the subsurface at temperatures much above 200°C, but gas does. Drilling to deep exploration targets during the next energy crisis will find gas, not oil.
3. Coal goes through a maturation process that is very similar to that of gas and oil. As shown in Figure 8–3, the maturation process for coal is a process whereby the fixed carbons are combined into complex sheet structures that create higher and higher orders of coal, with the simultaneous release of volatiles (v), mostly methane.

8.3.2 Types of kerogen

Four types of kerogen are known. They are types I, II, III, and IV as shown in Figure 8–4 and are characterized as follows:

Type I: High hydrogen-to-carbon ratio, characterized by normal and branched aliphatics, organic material is algal and amorphous, and is commonly found in lakes. This is the type of kerogen that occurs in oil shales, and though it is volumetrically rare, it is a very rich source rock for oil.

Type II: Intermediate H/C ratio, composed of napthenes, aromatics, and aliphatics, called herbaceous, commonly found in rocks deposited in marginal marine carbonates and shales. It is considered to be the volumetrically most important source rock for oil.

Type III: Low H/C ratio, composed of polycyclic aromatics and oxygenated functional groups, called woody, and is derived primarily from land plants. It is a good source rock for gas, but not oil. It is commonly found in association with coal.

Type IV: H/C ratio near 0.5, is derived from partially burned land plants. It is similar to charcoal and is extremely rare in nature.

Figure 8–4 shows that different types of kerogen can be identified in areas where the kerogen is immature. Through the maturation process

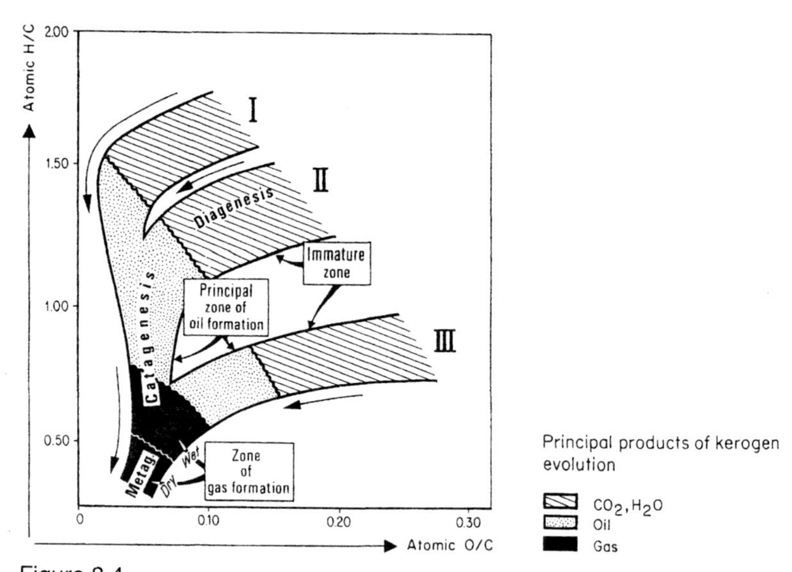

Figure 8.4
Van Krevelin diagram showing maturation paths for three types of kerogen. From Tissot & Welte (1978), reprinted by permission of Springer-Verlag.

the different kerogens successively lose their identities. In the oil window (during catagenesis, Fig. 8–3), natural gas and oil are expelled. Carbon remains and becomes fixed in complex ring structures. Continued maturation results in more fixing of carbon and successively lower H/C ratios that mask the original signature of the kerogen. By the metagenesis stage, it is not possible to identify the original kerogen types.

The principal means of identification of the composition of oils is through various types of chromatography. A typical gas chromatograph for a typical oil is shown in Figure 8–5. In this particular type of chromatograph, oil is dissolved in a solvent, the solvent is placed in the apparatus, and the apparatus is heated through time. As the boiling points of the various constituents are reached, bursts of vapors are given off and are recorded by the detector. Each burst on the chromatograph corresponds to the boiling point of one of the various constituents in the oil. For example, C_{15} represents the aliphatic $C_{15}H_{32}$, and so on. Identification of the various isomers (same chemical composition, but different structure), complex molecules, and different molecules with identical boiling points is extremely difficult. Fortunately, most oils are composed primarily of mixtures of aliphatics which can be identified quite readily by chromatography, and the isomers which are difficult to distinguish have physical and chemical properties that are normally similar to the basic aliphatic molecules.

8.3.3 Geochemical fossils

In some oilfields it is possible to identify different types of oil that have been derived from different source rocks. In some cases, usually when the source rock is relatively immature, it is possible to identify the source rock through geochemical fossils. Geochemical fossils include:

1. H/C ratio and O/C ratio as plotted on the Van Krevelan diagram (Fig. 8–4).
2. Carbon preference index.
   A. Most land plants have an odd carbon preference index. That is, they have more C_{15}, C_{17}, C_{19}, C_{21}, etc., than even-

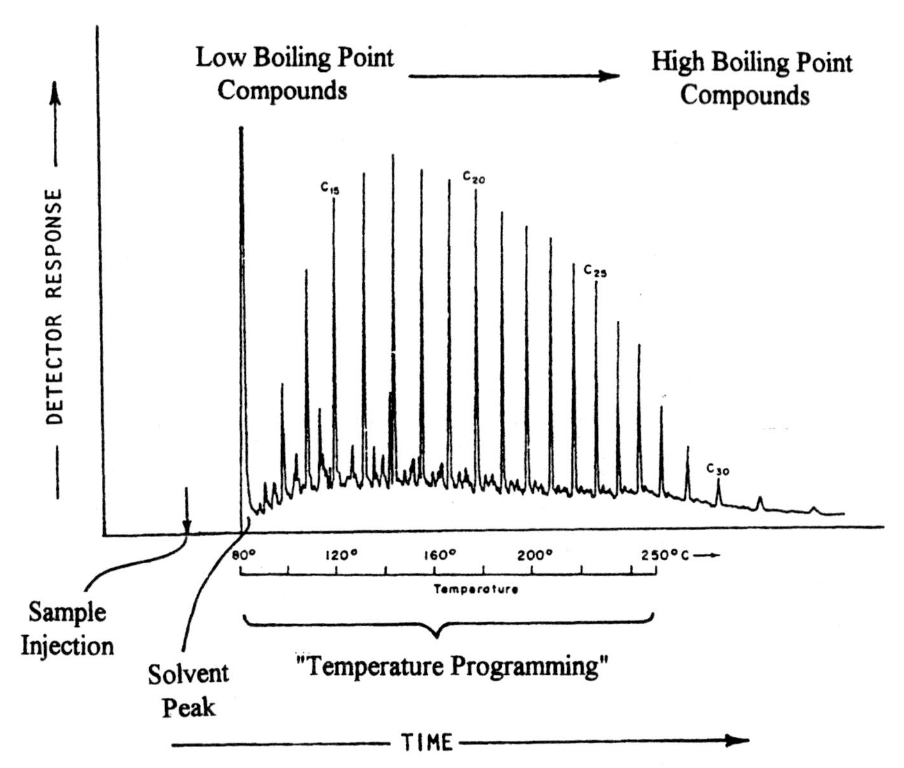

Figure 8.5
Typical gas chromatograph showing prominent normal aliphatic peaks labelled C_{15}, C_{20}, etc. From Barker (1982), reprinted by permission of AAPG.

numbered carbon chains. Most land plants have a large percentage of *n*-alkanes as waxes to prevent evapo-transpiration, but why they have an odd carbon preference is not known.

B. Many fatty acids derived from animals have an even carbon preference index.

3. The isoprenoids, pristane and phytane, are derived primarily from chlorophyll, which is derived primarily from algae. The chemical composition of chlorophyll is shown in Figure 8–6. In the subsurface the isoprenoid side chain tends to break from the porphyrin molecule at the oxygen link, and pristane (C_{19}) and phytane (C_{20}) show up on the gas chromatograph as peaks just below the C_{19} and C_{20} aliphatic peaks.

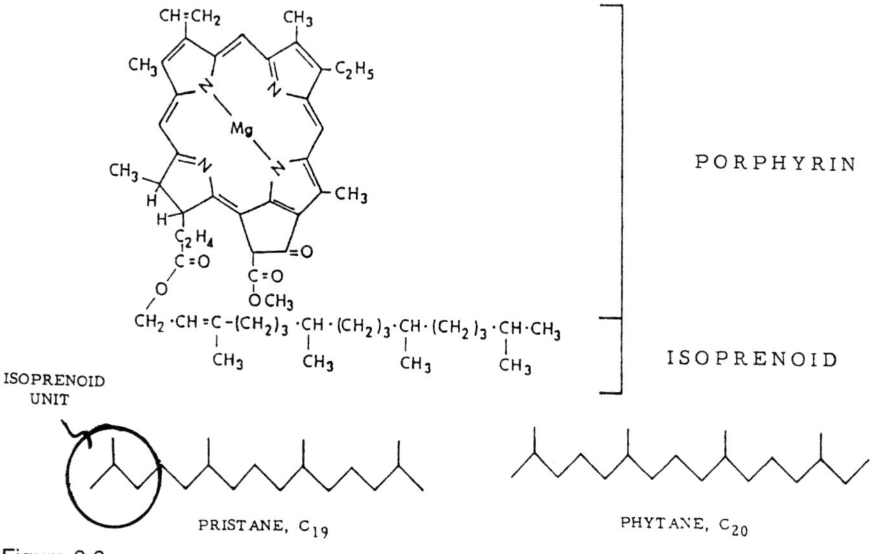

Figure 8.6

Chemical structure of chlorophyll, showing porphyrin nucleus and isoprenoid side chains, pristane and phytane. Modified from Barker (1982), reprinted by permission of AAPG.

These characteristics can normally be identified if the source rock is relatively immature and the oil has not been degraded. Identification of geochemical fossils becomes successively more difficult as the source rock becomes more mature (Fig. 8–7), or if the oil is degraded. Oil degradation can be caused by heat (cracking), biologic activity, or oxidation.

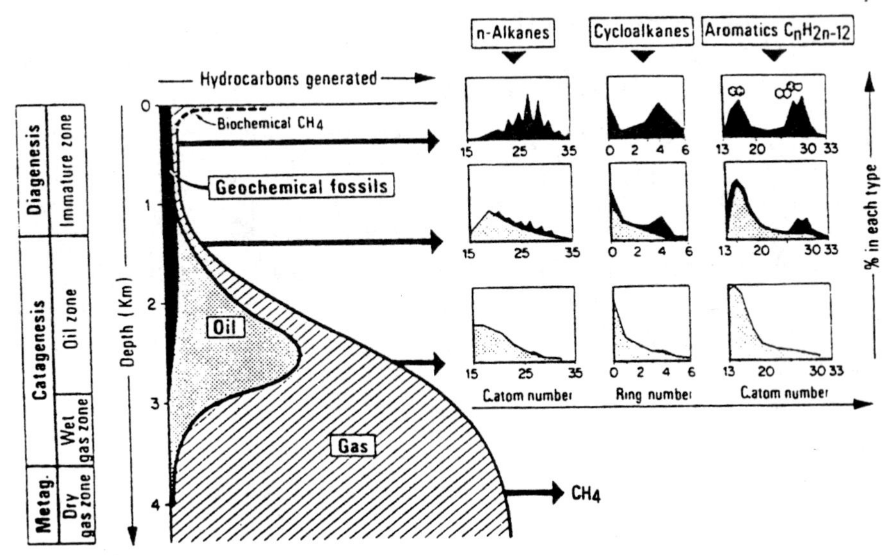

Figure 8.7
Generalized scheme of hydrocarbon formation showing evolution of hydrocarbon types as a function of burial depth and maturation. From Barker (1982), reprinted by permission of AAPG.

8.3.4 Biodegradation and water washing

A large number of bacteria exist that are able to metabolize hydrocarbons. The bacteria commonly live in oxygenated groundwater, and they selectively metabolize hydrocarbons with the highest potential energies, namely the short-chain *n*-alkanes, first. Thus, oil that has been exposed to water washing or biodegradation will be missing the shorter-chain *n*-alkanes. Such oils are very difficult to remove from a reservoir because they are gooey, tarry messes. They have very high viscosities and commonly form tar mats at the oil-water contact. This phenomenon is particularly common at oil-water contacts with fresh water.

8.4 Phase Relationships

With the exception of dry gas (pure methane) all oils and gases are composed of a *mixture* of hydrocarbons. These mixtures all have their own unique phase behaviors, because they are composed of different percentages of the component hydrocarbons.

Figure 8–8 includes four phase diagrams for different oils and gases. In all four figures the curves of paths A to B show the change that occurs as the fluid flows from the reservoir to the surface. The curves of

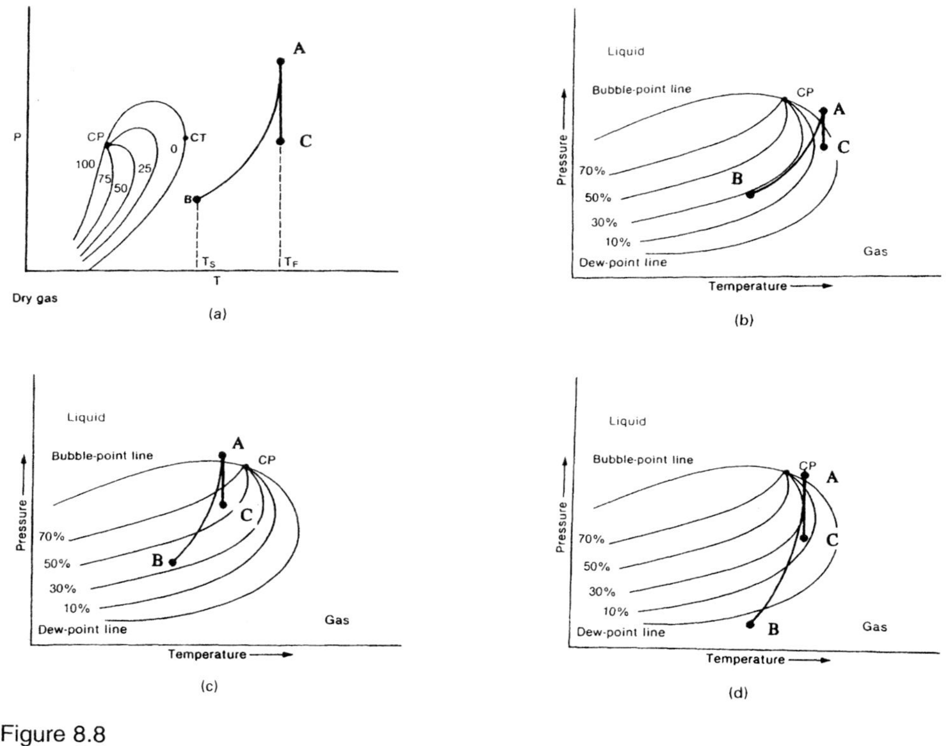

Figure 8.8

Phase diagrams. In all four parts, A —> B symbolizes change from reservoir conditions to surface conditions during production. Path A —> C symbolizes pressure, but not temperature, drop in the reservoir. (a) methane; (b) a typical crude oil; (c) a gas-condensate reservoir; (d) a typical retrogade gas-condensate reservoir.

paths A to C show the pressure changes that occur in the reservoir as fluids are produced through time. In the reservoir, pressure normally drops through time, but temperature does not. In the following discussion, we will discuss curves A to B, but realize that similar events can occur along the A to C pathway.

Figure 8–8(a) is a phase diagram for pure methane. As the gas moves from the reservoir to the surface, pressure and temperature change, but liquids are not produced because very low temperature (below –162°C) and/or pressure are required before methane will condense.

More typical phase diagrams for hydrocarbon mixtures are shown in Figure 8–8 (b, c, and d). The numbers on the diagram represent percent liquids. Thus, the 0 percent line represents the dew point for gas, the 100 percent line represents the bubble point for oil, and the critical point (CP) separates oil from gas in the reservoir.

Figure 8–8(b) shows a typical performance for an oil reservoir. As oil flows up the tubing (A to B) during production, gas will come out of solution, and such an oil well will have a certain producing gas-oil ratio that will help lift the oil to the surface. However, in the reservoir no gas will come out of solution until reservoir pressure drops to the bubble point line. When this happens, gas will come out of solution and a secondary gas cap may be formed. This may not be bad because gas cap expansion can be an effective drive mechanism. However, several precautions should be observed:

1. If this happens, it is important that the gas cap not be produced. If the gas cap is blown down, an important energy source for the oil could be lost and the oil recovery efficiency will be reduced.
2. If gas comes out of solution in the reservoir, three-phase flow will occur in the reservoir and the relative permeability of oil could be reduced.
3. As gas comes out of solution, the chemical composition of the oil changes. With the extraction of the lighter phases, the phase envelope will shift to the right, and the oil may become more viscous and difficult to produce.

Figure 8–8(c) shows typical behavior for a gas-condensate reservoir. Because A is to the right of the critical point (CP), this will start out as a gas reservoir. As pressure and temperature are reduced to the dew point line, condensate will start to drop out. This is acceptable in the tubing because the condensate is a very desirable product, but it is not good in the reservoir because the condensate will probably be lost to the reservoir as a residual. In such reservoirs it is extremely important to maintain reservoir pressure above or near the dew point line, or else condensate will be lost to the reservoir.

Figure 8–8(d) is a retrograde condensate gas reservoir. In progressing from point A to point C, notice that the condensate yield increases to 30 percent, then decreases back to zero. In such reservoirs, it may be possible to actually increase the condensate recovery by intentionally letting the reservoir pressure drop.

8.5 Formation Volume Factors

Consider two dry gas reservoirs that are identical in all aspects except that:

- Reservoir A is at 2,000 feet and pressure of 1,000 psi
- Reservoir B is as 10,000 feet and pressure of 5,000 psi

All aspects of the reservoirs are identical, including rock volume and porosity. Which reservoir has more gas in it? In the terms of the pore spaces of the rock available for gas storage, the volume is exactly the same. However, when the gas is decompressed to a common pressure base at the surface, reservoir B has approximately five times as much marketable gas as reservoir A because the gas in reservoir B is five times as compressed as the gas in reservoir A. Natural gases are compressible, and they expand on production to a common pressure base. Oil and water are compressible, but only very slightly.

Formation volume factors account for differences in oil and gas volumes due to pressure, temperature, compressibility, and phase changes during production. As used here and in the denominator of the equations in Table 3–1, all formation volume factors are expressed in formation units per surface units, such that, on inverting and multiplying, the units are changed to surface conditions.

As used here, FVF_{oil} will always be greater than 1 because oil always shrinks, and FVF_{gas} will almost always be less than 1 because gas always expands going from reservoir conditions to surface conditions. Gas flow lines at the surface are usually cold because of the pressure drop that occurs during transmission, and oil flow lines are usually warm because oil is still warm as it comes out of the ground.

Formation volume factors can be fairly tricky, and they are normally determined by reservoir engineers. Before introducing the most commonly used methods for determining formation volume factors, several cautions should be understood:

1. As reservoir conditions change during production, formation volume factors also change. The formation volume factor does not remain fixed throughout the history of a field.
2. For high condensate yield gas reservoirs, it is important that reservoir pressure not be allowed to drop below the dew point. If condensate is allowed to precipitate in the reservoir, a considerable amount of marketable condensate may be lost as residual oil to the reservoir.
3. A high *gas/oil ratio (GOR)* in an oil well could mean one of two things:
   (a) Gas is coming out of solution as the oil passes from reservoir conditions to surface conditions, or
   (b) The well could be producing from near the gas oil contact, and because of the pressure sink (or drawdown) near the well, gas from the gas cap is being pulled down (coned down) into the well.
4. All high-condensate yield gas wells, high GOR oil wells, and retrograde condensate gas wells should be handled with caution.

8.5.1 Oil Formation Volume Factor (FVF_o or B_o)

A satisfactory approximation of formation volume factor for oils below the bubble-point pressure may be obtained from Standing's (1952) correlation:

$$B_o = 0.9759 + 0.00012 \cdot X^{1.2}$$

where: $X = R_s\,(\gamma_g/\gamma_o)^{0.5} + 1.2 \cdot T$

where: R_s = solution gas/oil ratio in cubic feet/stock tank barrel
γ_g = gas specific gravity
γ_o = oil specific gravity
T = reservoir temperature in degrees F
γ_o = 141.5/(131.5 + API gravity)

Most normal crude oils have an API gravity near 25 and gas specific gravity near 0.7. Thus, for most normal crude oils X may be shortened to:

$$X = (R_s \cdot 0.882) + (1.2 \cdot T)$$

A typical oil reservoir will have an oil formation volume factor in the range of 1.1 to 1.5 reservoir barrels/surface barrels.

At depths where oil may be partially cracked, high GOR oil wells may be very difficult to distinguish from high condensate yield gas wells. Peculiar things happen in these types of reservoirs, and a reservoir engineer should be consulted before producing such reservoirs. For a more detailed discussion of formation volume factors, see Koederitz et al. (1989).

8.5.2 Gas formation volume factors (FVF_g or B_g)

There are several methods for calculating formation volume factors for gas. The most commonly used method is called the *pseudocritical method* and is a four-step process as follows:

Step 1: Calculate the *pseudocritical* pressure and temperature for the gas by the following equations:

$$T_{pc} = 167 + (316.67 \cdot \gamma_g)$$

$$P_{pc} = 702.5 - (50 \cdot \gamma_g)$$

Step 2: Calculate the *pseudoreduced* pressure and temperature by the following equations:

$$T_{pr} = (T + 460)/T_{pc}$$

$$P_{pr} = P/P_{pc}$$

where: T = reservoir temperature in degrees F
P = reservoir pressure in psia

Step 3: Estimate Z, the gas deviation factor, from Figure 8–9. Z is a factor used to express deviation from ideal gas behavior. As pressure increases linearly many petroleum gasses do not compress linearly. Many seem to overcompress, but at pressures above 5,000 psi they tend to undercompress. Additional corrections must also be made for impurities such as CO_2 and N_2, but these corrections are usually minor unless their content is high.

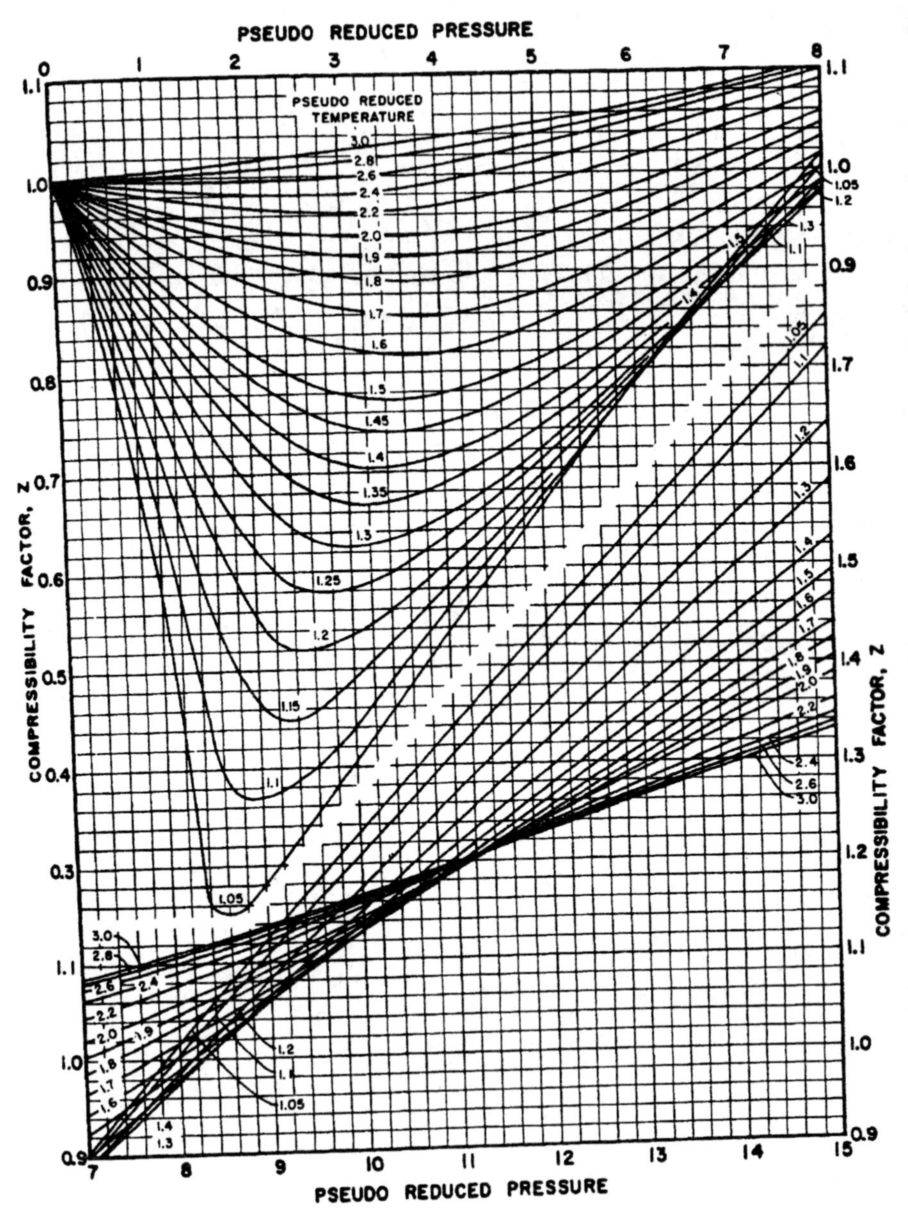

Figure 8.9
Z factor determination from pseudoreduced temperature and pseudoreduced pressure. From Standing, Marshall B., and Katz, Donald L., "Density of Natural Gases," Trans. AIME, 146, Fig. 2, p. 270. Copyright 1942 SPE, reprinted by permission of the Society of Petroleum Engineers.

Step 4: Calculate the gas formation volume from the following equation:

$$B_g = FVF_{gas} = .02829 \cdot Z \cdot (T + 460) / P$$

B_g units will be:

reservoir cubic feet/surface cubic feet

Gas formation volume factor check For dry gas reservoirs, reservoirs with low condensate yields (.65 gravity), gas formation volume factors may be grossly approximated by:

$$B_g = \text{Surface pressure/reservoir pressure}$$

where surface pressure is approximately 15 psi.

This should be used only as an order of magnitude check. Gas formation volume factors will normally be in the range of 0.01 to 0.001 reservoir cubic feet/surface cubic feet.

Example problems:

1. Calculate the formation volume factor for oil for a reservoir with the following:
   (a) Producing gas-oil ratio of 350 cubic feet/barrel
   (b) Gas gravity = 0.70
   (c) API gravity = 30°
   (d) Formation temperature of 200°F.

Solution

$$= 141.5/(131.5 + \text{API gravity}) = 141.5/(131.5 + 30) = 0.876$$

$$X = R_s \cdot (\gamma_g/\gamma_o)^{.5} + 1.2 \cdot T$$

$$= 350 \cdot (.7/.876)^{.5} + 1.2 \cdot 200 = 552.9$$

$$B_o = 0.9759 + .00012 \cdot X^{1.2}$$

$$= 0.9759 + .00012 \cdot 552^{1.2}$$

$$= 1.21 \text{ reservoir barrel/stock tank barrel}$$

2. Calculate and check the formation volume factor for gas for a reservoir having the following:
   (a) Gas gravity = 0.70
   (b) Reservoir temperature of 200°F.
   (c) Reservoir pressure of 3,500 psia.

Solution

Step 1:

$T_{pc} = 167 + (316.67 \cdot \gamma_g)$

$= 167 + (316.67 \cdot .7) = 388.7$

$P_{pc} = 702.5 - (50 \cdot \gamma_g)$

$= 702.5 - (50 \cdot .7) = 737.5$

Step 2:

$T_{pr} = (T + 460)/T_{pc}$

$= (200 + 460)/388.7 = 1.698$

$P_{pr} = P/Pc = 3500/737.5 = 4.746$

Step 3: Using Figure 8–9:

Enter P_{pr} (approx 4.75) from the top of the figure.

Find the P_{pr} intersection on the 1.698 (approx 1.7) line.

Read Z on the left side of the chart (0.87).

Step 4: Calculate B_g using:

$B_g = 0.02829 \cdot Z \cdot (T + 460)/P$

$= 0.02829 \cdot .87 \cdot (660)/3500$

$= 0.0046$ reservoir cubic feet/surface cubic feet

Ball park check: B_g = surface pressure / reservoir pressure.

$= 15$ psi/3500 psi $= .0043$

CHAPTER 9

Drive Mechanisms, Recovery Efficiencies, and Well Spacing

9.1 Types of Drive Mechanisms

There are two basic end-member types of drive mechanisms. A *depletion drive reservoir* is one in which the predominant producing mechanism is fluid expansion. A *water drive mechanism* occurs where a reservoir is open to an aquifer where reservoir pressure is maintained by incoming water. In reality, no reservoir behaves purely as either a depletion drive reservoir or a water drive reservoir; most are somewhere between. But the two end members serve as excellent models for understanding how reservoirs work, and reservoir behavior usually tends toward one or the other end member. Table 9–1 shows the different basic types of drive mechanisms and their associated primary recovery efficiencies.

Many so-called water drive reservoirs are not really open to artesian aquifer systems. Instead, such reservoirs are open to aquifers that are large enough that coefficients of expansion for water in the aquifer and enclosing rock cause the water in the aquifer to flow to the reservoir as though it were open to an artesian system. Water and rock have extremely low coefficients of expansion (2 to 10 × 10^{-6}/psi). Thus, for these factors to have an effect, the size of the aquifer must be very large (in fact infinite, to be true water drive). Even this system, which is called a water drive system (or partial water drive system), is really a rock-water-oil-gas expansion, depletion drive system.

Table 9–1 Classification of drive mechanisms and associated primary recovery efficiencies that can be expected.

Drive Mechanism	Recovery Efficiency (RE) (%)
I. Depletion drive	
A. Pure gas reservoir	60–80
B. Pure oil reservoir	5–30
1. Dead oil	5–10
2. Solution gas drive	10–30
C. Gas cap expansion for an oil reservoir	20–40
D. Gravity drainage	25–90
II. Partial water drive	
A. Pure gas reservoir	50–80
B. Pure oil reservoir	5–75
III. Water drive	
A. Pure gas reservoir	50–80
B. Pure oil reservoir	35–75
1. Edgewater drive (high dip, thin reservoir)	35–70
2. Bottom water drive (low dip, thick reservoir)	40–75

9.2 Drive Mechanisms and Recovery Efficiencies

Two important generalizations are: Water drive is best for oil recovery, and depletion drive is best for gas recovery.

The first generalization is not surprising, but the second may be. On first consideration, most people would guess that a water drive, which helps maintain reservoir pressure, would be better for both oil and gas. In fact, this is not so; depletion drive is commonly better for gas production. The simple analogy of an automobile tire helps. If you fill your tire with air to 32 psi, then take the valve stem out, it deflates exponentially (as in Fig. 9–1), and will eventually go completely flat. But the recovery efficiency is extremely high, on the order of 95 percent. Now if you fill the tire with oil to a pressure of 32 psi, then take the valve stem out, the oil will dribble out, the tire will go flat, and the recovery efficiency will be very low, perhaps on the order of 20 percent. The analogy works because the drive mechanism is the coefficient of expansion of the tire (rock) and oil. Oil and rock are just not very compressible compared to gas.

Now fill the tire with half gas and half oil, pump it up to 32 psi, and pull the valve stem out. What will happen? Clearly, it depends on where the valve stem is located. If it is the upper part of the tire, the gas cap will blow down and almost no oil will be recovered. Federal regulations require that associated gas reservoirs (gas associated with oil) may not be produced until the oil is produced. (Nonassociated gas reservoirs may be produced immediately.) If the valve stem is at the base of the tire, however, the gas cap will push the oil to the valve stem, and the oil recovery efficiency may be very high. Another important generalization is that: for a depletion drive reservoir, recovery efficiencies will be enhanced if the wells are placed in structurally low positions.

Now let's put oil and water, but no gas, in the tire, pump it up to 32 psi, and take out the valve stem. What happens? Again, it depends on the location of the valve stem. Let's put the valve stem at the top of the tire. What happens? Some oil is recovered, but not very much, because the expansion coefficients of oil, water, and rock (rubber) are low. Now let's start injecting water at the base of the tire to simulate a water drive reservoir (or a water flood). It should be fairly clear that for a water drive reservoir, recovery efficiencies will be enhanced if the wells are placed in structurally high positions.

Clearly, to place development wells properly in a reservoir it is important to know the drive mechanism. How does one know what type of drive mechanism to expect? This, again, is an area where the engineer

Figure 9.1
Classification of drive mechanisms and associated primary recovery efficiencies that can be expected.

and the geologist must communicate. The most certain way to know is to monitor the pressure as a function of production. If the pressure declines very rapidly, it is probably a depletion drive reservoir. If the pressure declines, then levels off or recovers, there is a good chance that the reservoir is a water drive or at least partial water drive reservoir. The problem is that by the time that pressure decline data is available, the wells have already been drilled.

In the development stage there must be some geologic input to determine the likelihood of the drive mechanism. There are no hard and fast rules, but generally:

1. For a reservoir to have a water drive:
   (a) It must be open laterally or downdip to a huge area (on the order of 100 square miles) if water expansion and rock expansion factors are to be invoked.
   (b) This clearly means that the reservoir rock must be a laterally continuous rock type such as a delta front sandstone, or a ramp carbonate. It cannot be a rock type that is typically discontinuous, such as a channel sand, a barrier island sandstone, or a carbonate reef.
   (c) The downdip area should be relatively free of faults.
2. Highly overpressured zones will normally contain depletion drive reservoirs.

Another way to predict drive mechanisms is to check production histories of nearby, comparable fields, if they are available. This method is not foolproof because local conditions such as faults or permeability barriers may isolate certain reservoirs.

9.3 Recovery Efficiencies

Clearly, recovery efficiencies are dependent on more than the drive mechanism. Recovery efficiencies are based on all of the following:

1. Reservoir quality (porosity, permeability, lateral and vertical continuity)
2. Drive mechanism
3. Well spacing
4. Time—how long will the wells be allowed to produce and at what rates?

5. Type of fluid recovered (gas, oil, heavy oil, tar)
6. Secondary and tertiary recovery techniques

Equations for recovery efficiencies have been developed, but in practice, most recovery efficiencies are *estimates* based on experience with all of the above taken into consideration.

Kh, permeability times thickness of the reservoir, is extremely important in the design of a field. Petroleum engineers spend a lot of time in the optimum design of recovery efficiency, number of wells drilled, cost, and effective drainage radius for each of the wells as a function of time. A brief and very tenuous summary of these is given in Table 9–2.

9.4 Well Spacing Design Criteria

Two of the most commonly asked questions of the development geologist and reservoir engineer are:

1. What is the best well spacing to use for a particular reservoir?
2. Given that well spacing, what is the expected recovery efficiency for the field?

Table 9–2 is based on experience in average reservoirs that have been developed over the last 50 years. The numbers are intentionally not precise because recovery efficiencies and well spacing are not independent variables, and each is truly dependent on many more variables than just drive mechanism and *kh*.

In the predevelopment stage, optimum well spacing should be determined by a full-blown economic analysis in which all things are considered, but the bottom line is present value profit (PVP) versus spacing, as shown in Figure 9–2.

Estimates of PVP are dependent on all of the following:

- cost of wells
- depth of wells
- number of wells
- recovery efficiencies, given a specific number of wells and well spacing
- reservoir characteristics, including porosity, permeability, fractures, anisotropy, permeability barriers, etc.

Table 9–2 "Rule of thumb" summary of KH, well spacing, and expected primary recovery efficiencies. The table is somewhat dangerous because recovery efficiencies are dependent on many more variables than these two. These are generalizations and the terms "extremely good, good, fair, and poor" are intentionally vague.

Water Drive Summary		
KH	Gas	Oil
Extremely good	70% RE, 320 acres	40%, 80 acres
Good	70% RE, 160 acres	35%, 40 acres
Fair	60% RE, 80 acres	30%, 20 acres
Poor	40% RE, 40 acres	20%, 10 acres
Depletion Drive Summary		
KH	Gas	Oil
Extremely good	80% RE, 320 acres	30%, 40 acres
Good	75% RE, 160 acres	20%, 30 acres
Fair	70% RE, 80 acres	10%, 20 acres
Poor	60% RE, 40 acres	5%, 10 acres

- drive mechanism
- price of the product
- time, inflation, flow rates
- secondary and tertiary recovery techniques etc.

The data in both Tables 9–1 and 9–2 give a ballpark indication of recovery efficiencies that can be expected in more or less average situations. It is important to keep in mind that every field is different and averages are dangerous.

9.5 Fractured Reservoirs

All bets are off when it comes to estimating recovery efficiencies in fractured reservoirs. Nelson (1983) recognized four principal fractured reservoir types:

Type 1: Fractures provide permeability and porosity for the field. Examples include Amal, Libya; some zones of La Paz/Mara, Venezuela; Big Sandy, Kentucky–Virginia.

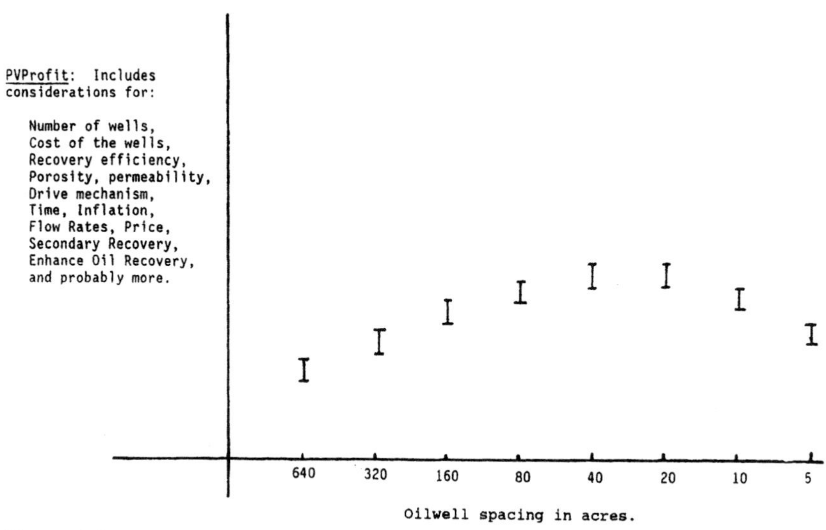

Figure 9.2
Well spacing versus Present Value Profit (PVP) to determine well spacing for a field. PVP is determined by a full economic projection (Chapter 11) where investment costs, which are primarily a function of the number of wells drilled (well spacing), are compared against future oil income, which is a function of recoverable reserves (in-place oil and recovery efficiency).

Type 2: Fractures provide permeability; matrix provides storage capacity. Examples include: Aga Jahri and Haft Kel, Iran; Spraberry, Texas; Altamont-Bluebell, Utah.

Type 3: Fractures assist permeability in an otherwise producible field. Examples include Kirkuk, Iraq; Gachsaran, Iran; Dukhan, Qatar; Lacq, France.

Type 4: Fractures provide an anisotropy that could significantly affect waterflooding; applies to most fields.

Fractures are common in all types of reservoir rocks. They are particularly common in anticlinal structures where the crest of the anticline has been extended through tension (Fig. 9–3). Fracture sets parallel to conjugate shear sets are also common in all types of compressional situations.

In most reservoirs, fractures do not constitute a major part of the porosity or storage capacity of the rock. It is rare for fracture porosity to exceed 1 percent of the total storage capacity for a reservoir rock. The most common situation is that hydrocarbons are stored in matrix poros-

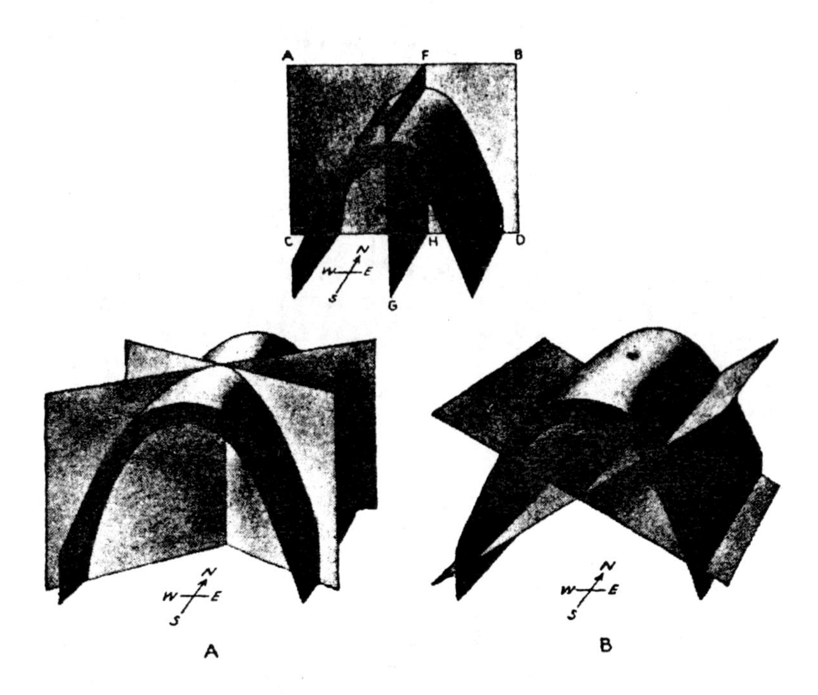

Figure 9.3
Common joint sets associated with conjugate shear sets in anticlines. From Billings (197s), pp. 168, 169, reprinted by permission.

ity and are then transmitted to the wellbore through fracture porosity. Clearly, under such circumstances the spacing of the fractures is one of the most important criteria for predicting reservoir performance.

For poor reservoir quality rocks that have significant vertical fracture patterns (Type 1, above), horizontal drilling is an important technique that helps open fracture permeability to the wellbore. Vertical wells through such reservoirs may produce spectacularly, or they may not produce at all. It is not uncommon in the Tamaulipas limestone of Mexico for individual wells to produce at rates of several thousands of barrels per day, while wells only 200 feet away hardly produce at all.

9.6 Horizontal Drilling

Horizontal drilling is a relatively new technology that has the potential to significantly increase recovery efficiencies in certain types of reservoirs. It has proven effective in:

1. Reservoirs with erratic or unpredictable horizontal permeability or in karsted areas.
2. Vertically fractured reservoirs, such as the Austin Chalk, where the fracture pattern is primarily in one direction. By drilling horizontally across (perpendicular to) the fracture direction, the effective drainage radius of a well can be significantly increased.
3. Thin oil columns where gas or water coning has been a problem. Horizontal wells expose more wellbore surface area to the reservoir and the pressure drawdown or sink at any one point is significantly lower in a horizontal well than in a vertical well.
4. Gravity drainage and low pressure reservoirs.
5. Production and injection wells for certain water flooding and enhanced oil recovery situations.
6. Situations where site preparation at the surface is extremely difficult or expensive.
7. Coal gassification. The quantity of gas given off by coal is largely a function of coal area exposed, and horizontal wells expose more reservoir area than vertical wells.

Horizontal drilling has not proved particularly effective in:

1. High permeability reservoirs and highly fractured reservoirs. These reservoirs are drained effectively by vertical wells anyway.
2. Thin beds that have little vertical permeability such as turbidites.
3. Homogeneous, poor quality reservoirs. Poor quality reservoirs remain poor whether drilled horizontally or vertically.
4. Vertically stacked reservoirs, such as on the side of a salt dome.

Horizontal wells are drilled using a variety of techniques. Most are drilled vertically to a kick-off point (KOP) located above the target horizon. The well is then deviated by any of a number of methods at rates that vary from as little as 1 degree per hundred feet to as much as 3.5 degrees per foot until the well is near horizontal (Fig. 9–4). A typical well might build its angle at 10 degrees per hundred feet, which means that a hole angle of 90 degrees is built over about 800 feet MD or less than 600 feet vertically and radially (Fig. 9–4). A tangent section (near horizontal section) is drilled (Fig. 9–5) until the target horizon is encountered, and then the well is deviated further (normally to horizontal) to stay within the target formation. Once in the formation, the drill bit tends to stay in the formation,

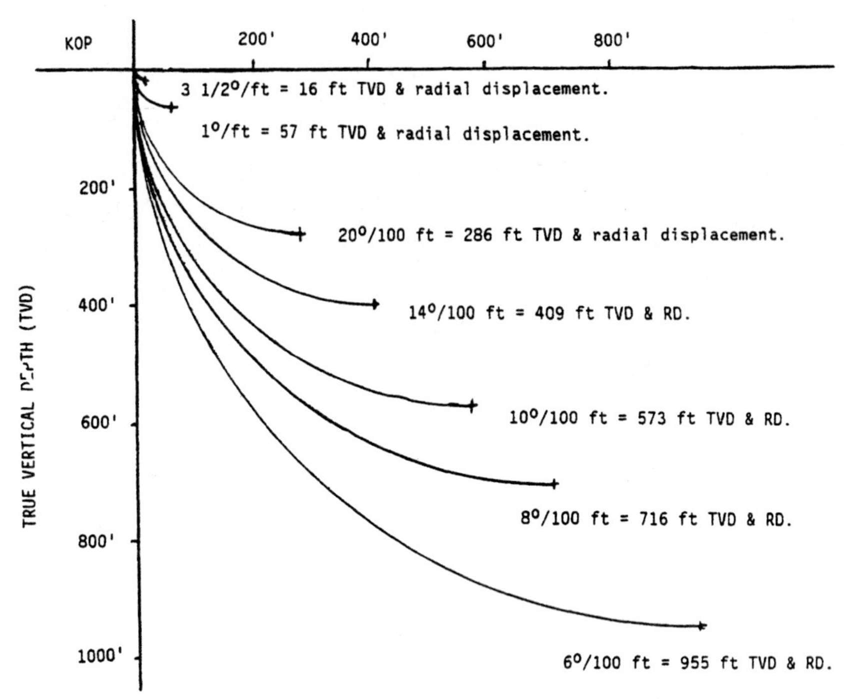

Figure 9.4
Radial displacement and true vertical depths of various turning rates for horizontal wells.

because most bits tend to slide off bedding planes that are encountered at very low angles, although porpoising (Fig. 9–5) can occur. The well is drilled for hundreds to sometimes thousands of feet horizontally.

Long radius wells (less than 10°/100 feet) may be drilled with conventional drill pipe and conventional directional drilling techniques (Fig. 9–6) such as:

1. **Whipstock:** A metal wedge inserted in the hole, to deviate the drill bit in hard rock situations.
2. **Bent sub:** A bent sub is a downhole drilling assembly where an offset bit is placed at the bottom of the hole. The bit is turned by a hydraulic motor that is powered by the pumped drilling mud. The drill pipe and sub are not rotated, such that the offset bit and sub move downhole in the direction of the offset bit. Bent subs are typically used in intermediate hardness rocks.

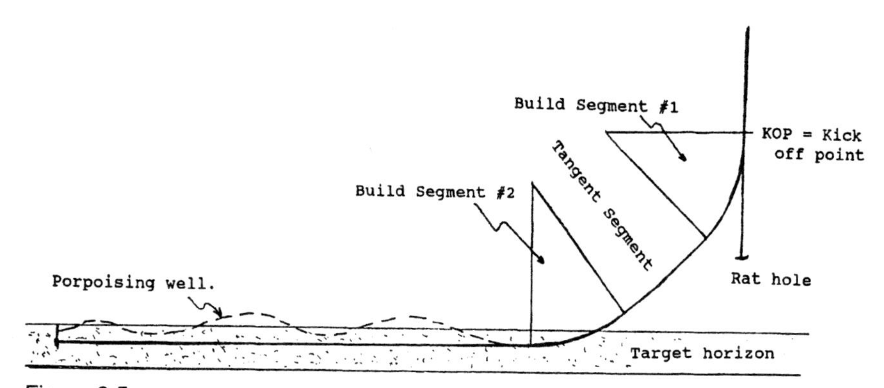

Figure 9.5
Typical well profile for proposed horizontal well (solid line) and actual, porpoising well (dashed line).

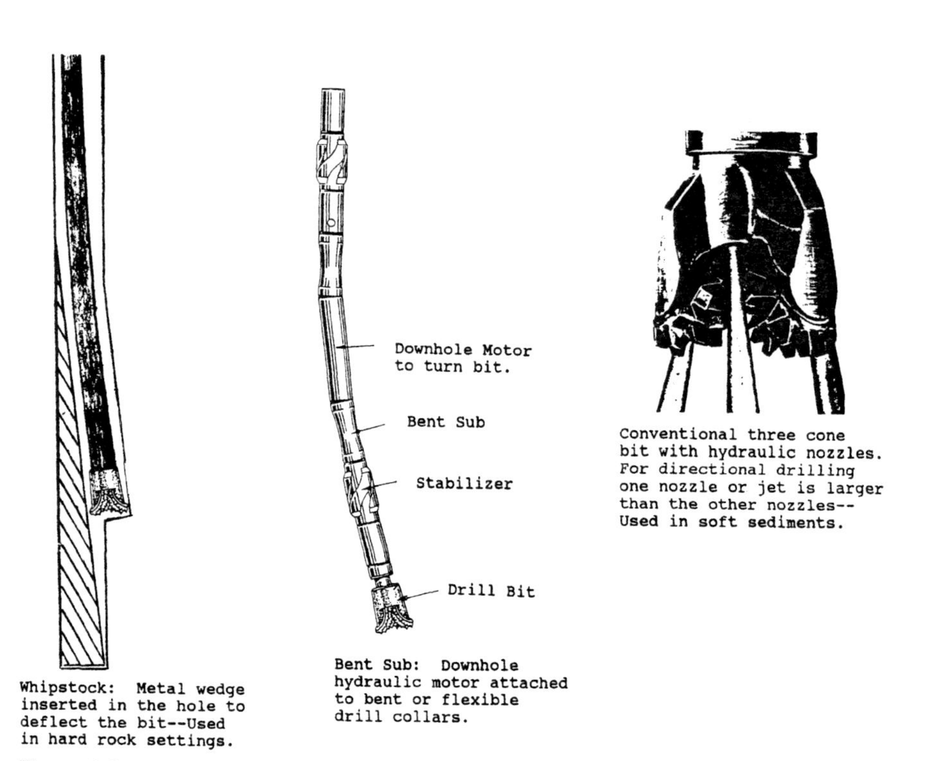

Figure 9.6
Methods used to drill long and intermediate radius segments of deviated holes.

3. **Jet drilling:** for soft rock situations such as the shallow Gulf Coast. In soft rock situations the large hydraulic nozzle (jet) on the drill bit may be oriented such that it will wash the hole in front of the bit and cause the well to deviate. Again, the drill pipe is not rotated while jetting.

For intermediate (3°/ft to 10°/100 feet) and short radius (1.5° to 3°/ft) holes, articulated drill collars are often used. For ultra-short radius holes (1.5°/ft or less), a vertical hole is normally drilled and underreamed, or widened, at the kickoff point. A special assembly is then run in the hole that commonly drills a horizontal hole either hydraulically or mechanically.

9.6.1 Disadvantages of horizontal wells

Logging Ultra-short and short radius wells cannot be logged. Intermediate and long radius wells can be logged using MWD (measurement while drilling) techniques, coiled tubing, or pumpdown logging equipment. All are more expensive and more likely to fail than conventional logs, but rapid improvements in MWD logging equipment are being made.

Geologic control Geologic control (by logs) in the horizontal segment of the hole is very poor. Control is often limited to knowing only whether the bit is in or out of the reservoir. If even a small unanticipated fault is encountered, it may be impossible to know where the bit is geologically.

Length of horizontal segment The length is primarily a function of radius of curvature. The tighter the corner and the longer the horizontal segment the greater will be the drag on the drill bit and drill pipe. In ultra-short and short radius wells it is difficult to drill horizontally beyond several hundred feet. In long radius wells, horizontal segments of over 5,000 feet have been drilled.

Hole conditions Rocks cave in, and the risk of hole collapse is always present. The longer the horizontal segment and the softer the rock, the greater the risk of hole collapse.

Cuttings Cuttings settle out in the horizontal segment of the well. Pump rates or drilling fluid viscosities must be increased to push cuttings out of the hole, and even then cuttings can be a serious problem.

Casing Casing cannot be run in ultra-short and short radius wells, and is very difficult to run in intermediate radius wells.

Completions Standard completions are difficult in horizontal wells. Acid pumped into horizontal wells commonly goes into the zones of highest permeability and is not distributed evenly. Sand control by resins or conventional gravel pack are impractical because of cost. Most completions are either open hole, perforated casing, or slotted liner completions. Prepacked gravel pack assemblies for use in slotted liners are utilized in soft-sediment areas such as the Gulf Coast.

Cores Cores may be taken without difficulty in long radius wells, and with more difficulty in intermediate radius holes. It is not possible to take cores in short or ultra-short radius wells.

Cement Cement is a problem in horizontal wells because both the pipe and the mud tend to favor the bottom of the hole. It may be difficult to get a thorough cement job above and below the pipe.

9.6.2 Horizontal well considerations

Horizontal drilling technology has advanced tremendously over the past ten years. Drilling costs have dropped markedly with experience, but horizontal wells still cost 15 to 250 percent more than conventional straight holes. They are much more difficult to drill, and the mechanical failure rate is much higher in horizontal wells than in straight holes.

When considering whether to drill a horizontal well, two questions should be asked:

1. What will be gained by drilling a horizontal well rather than using alternatives, such as drilling two vertical wells?
2. Is that gain worth the extra cost and risk?

The answers depend on the specific circumstances of the reservoir. In good reservoirs there is little evidence to date that recovery efficiencies are improved by horizontal drilling. Recovery rates may be different, but the overall recovery may not be improved by drilling horizontally. Figure 9–7 shows that for good reservoir rocks, two vertical wells capable of draining 20 acres (1,320 foot radius) will drain almost as much as one horizontal well with a 2,000 foot horizontal segment. If a horizontal well costs twice as much as a vertical well, there may be little reason to drill the horizontal well. The added risk of mechanical failure may make one

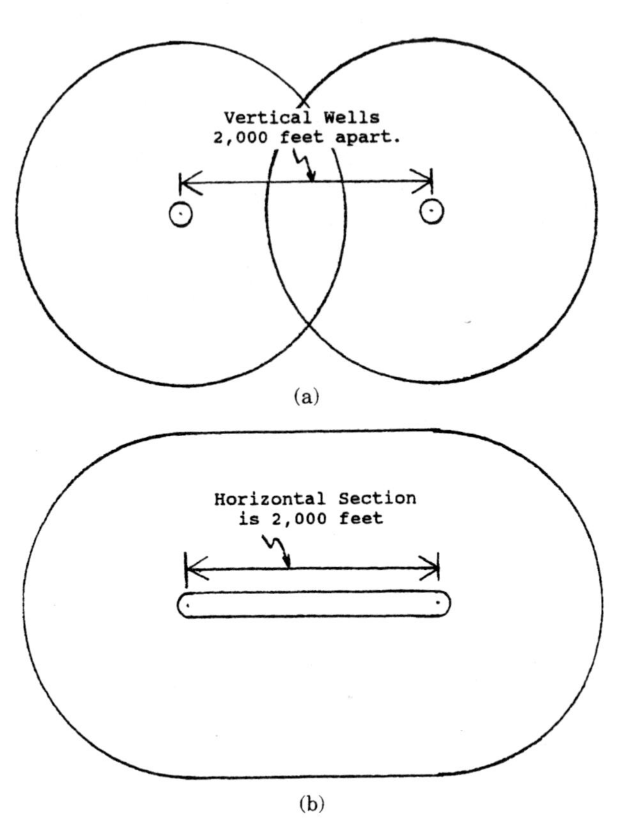

Figure 9.7
Comparison of effective drainage areas for (a) two vertical wells relative to (b) one horizontal well. All drainage points are assumed to have 1,320 foot drainage radius.

horizontal well much less attractive than two, or perhaps even three vertical wells.

On the other hand, for reservoirs with highly irregular horizontal permeability, ten vertical wells may not recover as much oil as one horizontal well.

CHAPTER 10

Volumetrics

The process of making a volumetric reserve estimate requires incorporating estimates using many different techniques into one composite estimate. It is important to realize that if all of the individual estimates are conservative, the resulting reserve estimate may be extremely conservative. If all estimates are optimistic, the resulting composite estimate may be outrageous.

10.1 Reliability of Data

10.1.1 Structure contour map

Reliability depends on the complexity of the field. Usually, when more wells have been drilled, the maps and reserve estimates are more reliable. However, in complicated fields, this is not necessarily true. Commonly, structure contour maps based on only one or two wells are simple and undisturbed. Each additional well adds some new complication, and after development, the geology is sometimes less certain than before development. This is not usually the case, but it can happen.

The structure contour map is the most important element in the estimation of reserves. Computers are very good at contouring simple, unfaulted structures, but by themselves, computers simply do not handle faults well. No structure contour map should ever be finalized without careful scrutiny by a geologist.

10.1.2 Isopach map

Reliability is as good as the structure contour map, the paycounts, and the development geologist. Once fluid levels have been identified, paycounts made, and structure contour maps prepared, there should be little room for interpretation in making the isopach. The preparation of isopachs should be mechanical once the above interpretations have been completed.

Again, computers are good at calculating volumes above or below particular horizons on a contour map, but they are not good at generating isopach maps. If a computer is used to generate isopach maps, it should not simply contour net or gross thickness estimates from petrophysics. Properly prepared gas and oil isopach maps require that structure, stratigraphy, and fluid levels be honored rigorously through the use of fault plane maps, structure contour maps of reservoir horizons, net reservoir thickness isopachs, and the intersection of fluid levels with those maps.

10.1.3 Porosity, fluid saturation, and paycount

Reliability is based on the quality of the logs and the well-log analyst. In analyzing identical data, different well-log analysts will almost invariably arrive at different conclusions. Fortunately, the differences should be offsetting if the analyses are honest. That is, if a particular analysis has an unusually optimistic paycount, poorer quality reservoir rocks may have been included in the paycount. The optimistic paycount should be offset by

- lower porosity estimates,
- higher water saturation estimates
- lower recovery efficiency estimates

10.1.4 Formation volume factor

Reliability is based primarily on knowledge of the composition of the reservoir fluids, plus formation temperature and pressure. If the composition of the gas or oil is not known, or if reservoir temperature and pressure are not known, these data can be estimated from regional geologic trends, but their reliability may be poor.

10.1.5 Recovery efficiency

Engineering techniques have been devised for estimating recovery efficiencies, but recovery efficiencies are based on such a large number of variables that analogs and experience are still considered the best way to estimate recovery efficiencies.

10.1.6 Data summary

The components of reserve estimates come from different types of data and from different people. Geologic maps are interpretations. Estimates of porosity, fluid saturation, paycounts, and recovery efficiencies are engineering estimates and are subject to interpretation.

It is possible that a worst-case scenario or best-case scenario is desired, but in most cases it is important that the engineer and geologist communicate with each other to make certain that each is making a best-estimate or most-likely interpretation of the data. Again, if any one data set or interpretation is wrong, the final results will be wrong.

10.2 Predevelopment Studies

In this complicated business, communication and coordination of many people with diverse backgrounds can be helped, before the development drilling rig arrives on location, by a predevelopment study. A study is particularly appropriate for offshore situations with a time gap between exploratory drilling and development drilling. In such a predevelopment study, logically coordinated by the development geologist and/or the reservoir engineer, the coordinator summarizes what is known about the field and informs all those involved as to approximate scheduling. The writer asks for input from all specialists and compiles it all together in a report. Archaeological surveys or environmental impact statements might be included, if available, and any unusual or potentially dangerous situations should always be highlighted. All likely development wells should be shown on the development geologist's map hypothetically drilled, and as many contingencies as possible should be considered.

It is important that poorly understood areas are drilled early so that development plans can be adjusted. The predevelopment study is a good place to warn managers that some high-risk wells are going to be drilled early. Managers don't like high risk and uncertainty; they want "sure shots" because they want to book reserves as quickly as possible. As an example, for a 24-well development program, a sequence something like the following should be prepared:

1. **Well #1—Sure shot:** Neither you nor the manager wants a dry hole on the first well.

2. **Well #2—Reasonably sure shot:** but test a new area. In order to develop three-dimensional control on structure (particularly faults) and stratigraphy it is important that exploratory wells not be duplicated exactly. In cases where the development wells are exact duplicates of exploratory wells, no additional geologic information is obtained. In one offshore Louisiana case, four development wells exactly duplicated the exploratory wells. Fault cuts were observed in all wells, but because the development wells were exact duplicates it was not possible to determine the orientation of the faults. Had the development wells been slightly offset from the exploratory wells, a great deal of information on the orientation of the faults could have been obtained.
3. **Well #3—Wildcat:** Go to the geologically unpredictable area, especially if it has large potential. Significant new geologic information could change the entire development plan. Don't let your manager put this well off.
4. **Well #4—Sure shot:** Your manager is a little nervous about #3. He/she needs some reassurance that you are not out of your mind.
5. **Well #5—Good shot with plans to go deep:** Casing is one of the most expensive parts of any well. Since you must run production casing in a sure shot well anyway, plan to drill out of the production casing and explore for deep targets. Again, this should be done early in the development program because if you find something big, the development plans could change significantly.

Finally, managers don't drill wells for information. They drill for oil and gas reserves based on sound economics. The development geologist must attempt to design wells, including the order of the wells, such that drilling can produce important three-dimensional information early in the development program, while simultaneously seeking oil and gas reserves based on sound economics.

CHAPTER 11

Economic Projections and Decline Curves

After a discovery has been made, a company has a number of options. One option is to sell the field to another company before it has been developed. Many companies do not like to do this because commonly the reserves are not well-defined immediately after discovery, and hence the value of the field is uncertain. Most companies prefer to develop their own discoveries because additional reserves are commonly added as development drilling proceeds.

The geologist who develops the field must be familiar with all the options if the field is to be of maximum profit to the company. In large oil companies, economic projections are normally developed by reservoir engineers. However, in small companies the development geologist may perform the projection, and even in large companies, it is important that the development geologist have a working knowledge of economics and price projections because decisions are made based on bottom-line economics, not geology.

This chapter is divided into four principal parts. The first consists of definitions of common terminology used throughout the petroleum industry.

The second is devoted to the time value of money. The petroleum industry is characterized by large front-end capital investments, followed by production and income that trickle in over the next 20 to 100 years. In order to make good investment decisions, it is important that the time value of money be considered so that profit-to-investment ratios can be properly evaluated.

The third section deals with decline curves. Production from oil and gas wells declines through time, which means that future income declines through time. Decline curves are included at this point in the book because the mathematical equations for exponential decline curves are almost identical to those for the time value for money.

The chapter concludes with a practical example.

11.1 Some Important Definitions

11.1.1 General petroleum terminology

- **Dry hole cost (DHC):** Estimated cost to drill to a target; does not include completion of the well. It may or may not include lease costs.
- **Lease costs:** Cost to lease the land either from the landowner or the government. This normally includes payment of a bonus plus annual rentals.
- **Completion cost (CC):** Cost to run casing, production tubing, downhole equipment, wellhead equipment, and peripherals which make the well physically capable of producing. It may or may not include pipeline costs and collection facilities such as tank batteries.
- **Completed well cost (CWC):** Dry hole cost plus completion cost.
- **Investment:** Money spent on a venture of any sort.
- **Income:** Money returned as a result of an investment.
- **Gross:** Total income (100 percent) of the income or production.
- **Net:** Investors' percentage of the income or production.
- **Working interest:** Percentage of all costs to be paid on an investment. A working-interest owner has agreed to pay a certain percentage of all costs incurred in the drilling of a well. A working-interest owner is also liable for any problems, either economic or environmental, that may result from the venture.
- **Net revenue interest:** Percentage of income to be paid to working-interest owners. Normally, a net revenue interest is the same as the working interest after the proportional share of royalties and overrides have been taken out.
- **Overriding royalty interest (override or ride):** Off-the-top percentage of any income. A royalty interest owner has no financial obligations towards the drilling of the well or field.
- **Profit:** Income minus investment. It may be gross or net. A true profit occurs only when income is greater than investment.

- **Undiscounted profit:** Same as profit. That is, no risk discounts or present value discounts have been applied to the income or investment.
- **Profit-to-investment ratio:** Profit divided by investment, usually expressed as a percent.
- **Writeoff:** An investment that may be subtracted from income before paying income taxes. One hundred percent of a dry hole may be written off.
- **Capitalized item:** Capitalized items are normally real items with real value that diminishes or depreciates through time. For example, wellhead equipment on a completed well is normally capitalized. Capitalized items may be written off through time using a depreciation schedule.

11.1.2 Time value of money and present value discounts

Interest and present value equations are summarized in Table 11–1.

- **Inflation:** The increase in the cost of buying goods as a function of time expressed as a percent. It may be thought of as the deflation of the buying power of the dollar.
- **Present value:** Adjustment of any monetary income or expenditure to an "as of" date. The adjustment is for the changing value of the dollar as a function of time.
- **As of date:** Any arbitrary date. Commonly January 1 of the present year or the year in which an initial investment is made. All present value calculations are adjusted to this date.
- **Present value profit (PVP):** Present value of all income minus present value of all investments.
- **Present value profit after tax (PVPAT):** Present value of all income minus present value of all investments, all adjusted for the tax structure of the individual company.

11.1.3 Risk

- **Probability of success (Ps):** Statistical probability that a given venture will make a profit.
- **Risk:** Same as Ps. As an example, if Ps = .2, it is the same as saying that the investment has a 20 percent (one in five) chance of success.
- **Risk capacity (RC):** (PVP of the whole venture – DHC)/DHC. A comparison of the expected PVP of the whole venture against the

Table 11–1 Summary of interest and present value equations.

- **Interest:**
  1. As used in text, interest is the percentage of the income or investment assigned to a company or individual.
  2. As used here it is the payment, usually expressed as a percentage, for the use of borrowed money.

- **Effective interest** (i): Same as simple interest. That is, interest accumulated per certain length of time, usually one year.

$$Yn = C\,(1 + i)^{n} \quad \text{(Interest equation)}$$

or:

$$C = Yn\,(1 + i)^{-n} \quad \text{(Present value equation)}$$

where: Yn = amount accumulated after n time periods
C = initial investment
i = simple interest rate
n = number of time periods (usually years)

Note: i in years is not equal to 365 times i in days

- **Nominal interest** (j): Interest accrued through continuous compounding, sometimes referred to as instantaneous compounding.

$$Yn = C\,e^{nj} \quad \text{(Interest equation)}$$

or:

$$C = Yn\,e^{-nj} \quad \text{(Present value equation)}$$

where: Yn = amount accumulated after n time periods
C = initial investment
n = number of time periods
j = nominal interest rate

Note: j in years is equal to 365 times j in days

And: $i = e^{j} - 1$ and $j = ln(1 + i)$

- **Doubling time:** Amount of time that it takes for an investment to double. For example, at 10 percent, an investment should double in value every seven years. Conversely, at 7 percent, an investment should double in value every 10 years, not counting inflation.

- **Rate of return (ROR):** Interest rate that is equivalent to setting the present value profit for the investment to zero. This is one of the very best profitability measures because it compares a particular investment to a theoretically equivalent interest rate, but it is difficult to solve. It must be solved incrementally or graphically.

risked investment. It is an index of the amount of risk that a particular venture can withstand. RC · Ps should be considerably greater than 1 before a venture can be considered feasible.

- **Risk discounted PVP:** (Present value of all income times Ps), divided by the investment.
- **Payout time:** Time from the investment until income equals investment. It can be very sensitive to present value assumptions.
- **Reversionary interest:** Net revenue interest reverts or changes at some point in time, commonly after payout.
- **Upside potential:** Potential that reserves and income are greater than expressed.
- **Downside potential:** Potential that reserves and income are less than expressed.

11.2 Time Value of Money

A summary of interest and present value equations is given in Table 11–1.

11.2.1 Effective interest

When you borrow money from the bank, you must pay interest on that loan. Conversely, if you loan money to someone, they should pay interest to you. Simple interest, commonly referred to as effective interest, is a percentage of the money loaned per length of time, usually one year. For example, the value of $1000 invested at 10 percent effective interest over one year, but compounded over different time periods, is given by the equation

$Yn = C \cdot (1 + i)^n$ and is explained in Tables 11–1 and 11–2.

Table 11–2 Interest compounded on $1000.

	Yn	*C*	*i*	*n*
Compounded annually	$1100.00	$1000	.10	1
Compounded monthly	$1104.71	$1000	.10/12	12
Compounded weekly	$1105.06	$1000	.10/52	52
Compounded daily	$1105.15	$1000	.10/365	365
Compounded hourly	$1105.17	$1000	.10/8760	8760
Compounded instantaneously	$1105.1709	$1000	$j = .10$	1

11.2.2 Nominal interest

Table 11–2 illustrates the difference between effective compounding (over discrete time intervals) and nominal (instantaneous) compounding. By convention, effective interest rates are designated using the letter "*i*" and nominal rates using the letter "*j*." Bankers learned long ago that instantaneous compounding gives a higher return on an investment, and thus, most interest rates today are nominal. The distinction between effective and nominal interest is also important in decline curves.

From the table, it is apparent that a 10 percent interest rate compounded annually does not give the same yield as 10 percent divided by 12 and compounded monthly. That is:

$$i_{\text{annually}} \neq 12 \cdot i_{\text{monthly}} \neq 365 \cdot i_{\text{daily}}$$

Instead,

$$i_{\text{annually}} = i_{\text{monthly}}^{12} = i_{\text{daily}}^{365}$$

One of the conveniences of nominal interest rates is that:

$$j_{\text{annually}} = 12 \cdot j_{\text{monthly}} = 365 \cdot j_{\text{daily}}$$

Inflation can be thought of as the reverse of interest. In Table 11–1, it should be apparent that the present value equations are derived from a simple rearrangement of the interest expansion equations.

The concept of inflation is important to the petroleum industry because investments are usually made on one date, and income from that venture occurs over an extended period of time in the future. The present value equations adjust all investments and future income to a common "as of" date so that investments made on one date can be compared against income received on another date.

Example problems:

1. Calculate how a $1,000 investment will grow over seven years at an *effective* interest rate of 10 percent.

Solution:

$$Yn = C\,(1 + i)^n = \$1{,}00 \cdot (1.10)^7 = \$1{,}949$$

2. Calculate how a $1,000 investment will grow over seven years at a *nominal* interest rate of 10 percent.

Solution:

$$Yn = C\,e^{nj} = \$1{,}000 \cdot e^{(7 \cdot .10)} = \$2{,}014$$

3. Convert the effective interest rate of 10 percent to a nominal rate.

Solution:

$$j = ln(1 + i) = ln(1.10) = .0953$$

4. If you receive an income of $20,000 ten years from today, what is the equivalent buying power of that $20,000 in today's dollars given that inflation is 7 percent *effective*.

Solution:

This question is asking for the present value of $20,000 seven years from today.

$$C = Yn\,(1 + i)^{-n} = \$2{,}000 \cdot (1.07)^{10} = \$10{,}167$$

This is saying that $20,000 ten years from now will buy the same as $10,167 today.

5. If you receive an income of $20,000 ten years from today, what is the equivalent buying power of that $20,000 in today's dollars given that inflation is 7 percent *nominal*.

Solution:

$$C = Yn\,e^{-nj} = \$20{,}000 \cdot e^{(-10 \cdot .07)} = \$9{,}932$$

11.3 Decline Curves

11.3.1 Hyperbolic decline

Decline curves are one of the simplest forms of material balance that reservoir engineers perform. Most individual wells and groups of wells producing from a reservoir, from a field, or even from a nation will decline along a curve that can be modeled using some sort of hyperbolic decline curve. The hyperbolic decline curve is defined by:

$$q_t = q_0 / [(1 + (a_0 \cdot t/h))^h]$$

where:

q_t = flow rate at any given time
q_0 = initial production rate
a_0 = initial nominal decline rate
t = producing time
h = hyperbolic factor

where: $h = 1$ = harmonic decline
$h = 2$ and greater = hyperbolic decline (analyzed with type curves)
h = infinite = exponential decline curve

Note that as h goes to infinity, a special case of the hyperbolic decline curve, the exponential decline equation, is generated:

$$q_t = q_o \cdot e^{(-a \cdot t)}$$

While the exponential decline curve does not ever simulate any reservoir perfectly, it has some special properties that make it very easy to use (especially on a spreadsheet), and the concept is fundamental to all reservoirs. It is the model against which most reservoirs around the world are compared.

11.3.2 Exponential decline curve

For illustration purposes, Figure 11–1 shows how the flow rate of an oilwell will decline through time if it comes on line at 100 barrels per day and experiences a 30 percent per year effective decline rate. At the end of successive years, the well will flow at:

Initial flow rate:	100 barrels of oil per day (BOPD)
After 1 year:	100 BOPD · (1 – 0.3) = 70 BOPD
After 2 years:	70 BOPD · (1 – 0.3) = 49 BOPD
After 3 years:	49 BOPD · (1 – 0.3) = 34 BOPD
.	
.	
.	
After 12 years:	1.9 BOPD · (1 – 0.3) = 1.4 BOPD

Notice that production never goes to zero. In the tail end of the decline curve, the curve is asymptotic to the zero production line.

If this data is plotted on semilog paper (Fig. 11–2), the decline curve becomes a straight line, which again, does not go to zero because semilog paper has no zero line. In actual practice, some costs associated with all oil or gas production may increase with time. For instance, as oil production declines, water production usually increases, and there is a cost for the disposal of that water. Because of these costs, all wells become uneconomic before production reaches zero.

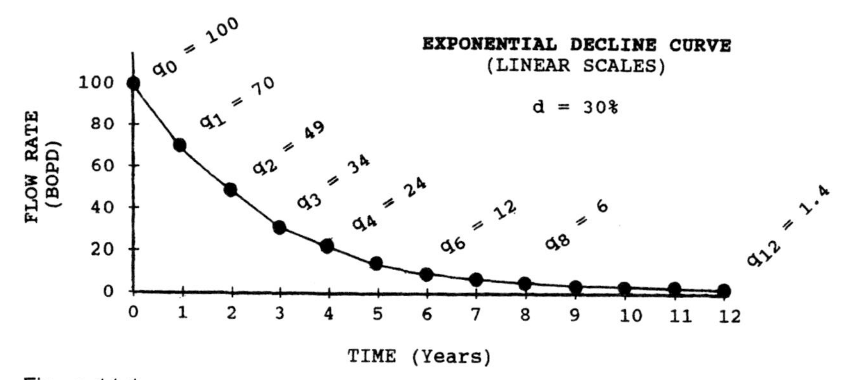

Figure 11.1
Production rate on a linear scale versus time for a 30 percent effective decline rate.

11.3.3 Economic limit

Every well and every field have an operational cost. The economic limit for the well or field is reached when operation and disposal costs exceed income from the well.

11.3.4 Reserves from decline curves

Exponential decline rate curves plotted on semilog paper are very powerful because:

1. They represent a visual model of the performance of the well or group of wells.

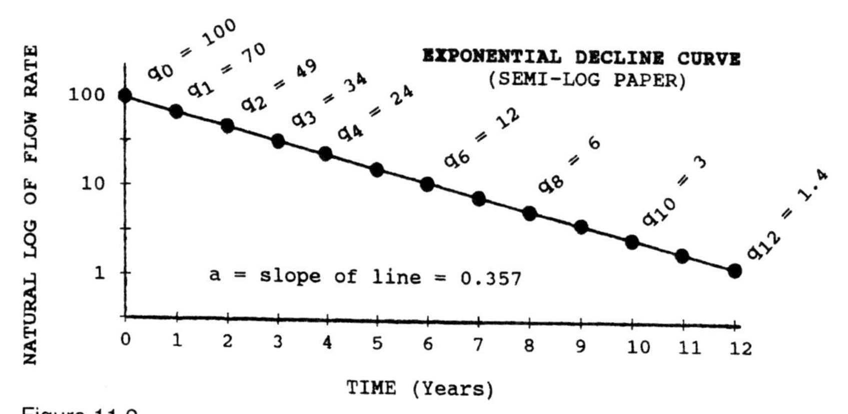

Figure 11.2
Natural log of production rate (semilog paper) versus time for a 30 percent effective or 35.7 percent nomial decline rate.

2. Departures from the curve commonly represent some physical phenomenon that is occurring in the reservoir. Departures can be seen visually and noted.
3. Unusually high decline rates may be the signature of a depletion drive in an oil reservoir.
4. By far the most powerful aspect of decline curves is that reserve estimates can be calculated quickly and easily once the decline rate has been established.

Table 11–3 summarizes the decline rate equations. Of interest is the *reserves* equation:

$$Q = (q_0 - q_{ec}) / a$$

where:

Q = reserves
q_0 = initial flow rate
q_{ec} = flow rate at the economic limit
a = nominal decline rate

This equation is important for several reasons. The first is that, as already stated, reserves can be calculated quickly and easily once q_0, q_{ec} and a have been established. When a well comes on line, reservoir engineers plot the production rates within several months after first production in an attempt to establish the decline rate. If the reserves calculated from this equation do not match volumetrics, engineers and managers will want to know why. In some cases it may be possible for these calculations to help the geologist redefine reservoirs where uncertainty exists. In some cases, discrepancies may indicate that faults are leaking or that permeability barriers exist that the geologist does not know about.

The second reason is that the worldwide average nominal decline rate (a) has been estimated to be 30 percent. Because q_{ec} is usually small and relatively insignificant in the equation, a quick and easy way to estimate reserves is by the following approximation:

$$Q = (q_0 - \text{almost } 0) / 0.3$$

or $Q = q_0$ times 3 (in years)

or Q = three years at the current rate

Throughout the consulting industry, this is a commonly used algorithm for estimating reserves in areas where little is known about the geology or about the well. It tends to be conservative because the worldwide decline rate is probably somewhat less than 30 percent.

Table 11–3 Summary of exponential decline rate questions.

- **Effective decline rate:**

$$d = (q_t - q_{t+1}) / q_t$$

where: d = effective decline rate

d in years is *not equal* to 12 · d in months nor 365 · d in days

q_t = flow rate at any point in time

q_{t+1} = flow rate at some later point in time

- **Conversion from effective decline rate to nominal decline rate:**

$$a = -ln(1 - d) \qquad d = 1 - e^{-a}$$

or: $a = (ln\ q_t - ln\ q_{t+1}) / t$

where: a = nominal decline rate

d = effective decline rate

Note: time units (years, months, or days) must remain the same.

and a in years is equal to 12 · a in months or 365 · a in days.

- **Exponential decline curve equations:**

$$q_t = q_0\ e^{-at}$$

where: q_t = flow rate at any point in time

q_0 = initial flow rate

a = nominal decline rate

t = time

Note: all time units must be consistent (years, months, or days). Units cannot be mixed.

- **Reserves:**

$$Q = (q_0 - q_{ec}) / a$$

where: Q = reserves

q_0 = initial flow rate, or flow rate at any point in time.

q_{ec} = economic limit flow rate equivalent.

a = nominal decline rate

Note: again, all time units must be consistent

- **Time to economic limit:**

$$t = (ln\ q_0 - ln\ q_{ec}) / a$$

Note: again, units must be consistent (years, months, or days)

In Table 11–3, the difference between *effective* and *nominal* or *instantaneous* decline rates is significant. The distinction is similar to that already expressed in the discussion of effective and nominal interest rates. Effective refers to flow rate changes that occur over discrete time intervals, whereas nominal refers to instantaneous changes expressed over certain time periods. It is recommended that effective decline rates be converted to nominal and that the nominal decline rate equations be used so that daily, monthly, and annual rates can be converted by simple multiplication and division. Also, exponents can be added or subtracted when decline rate equations are combined with present value equations.

A third reason that the equation is important is that q_0 does not need to be the initial flow rate. The initial flow rate, q_0, may be from any time forward. In other words, "three years at the current rate" is a reasonable estimate of the remaining reserves from any time forward, especially if no decline rate has been determined.

It has been observed that, for the early stages of production, most reservoirs can be simulated fairly accurately using exponential decline curves. Towards the end of production, most reservoirs deviate from the curve. In poorly consolidated sediments such as the Gulf Coast, production rates commonly fall below the exponential decline rate. In hard rock situations with good water drives, production rates commonly exceed the exponential decline predictions. Reservoir engineers adjust their predictions by using harmonic, hyperbolic, double exponential rates, or other simulation techniques.

Example problems:

1. A gas well comes on line flowing 700 mcf/day. Exactly one year later the well is flowing 500 mcf/day. What is the effective decline rate for this well?

Solution:

$$d = (q_t - q_{t+1}) / q_t = (700 - 510)/700 = 0.2714$$

2. Convert an effective decline rate of 27.14 percent to a nominal rate.

Solution:

$$a = -ln(1 - d) = -ln(1 - .2714) = .3166$$

3. Estimate the flow rate for the well after five years of production.

Solution:

$$q_t = q_0 e^{-at} = 700 \cdot e^{(-.3166 \cdot 5)} = 144 \text{ mcf/day.}$$

4. If the economic limit is considered to be 10 mcf/day, calculate the time required to reach the economic limit.

Solution:

$$t = (ln\ q_0 - ln\ q_{ec})\ /a = (ln\ 700 - ln\ 10)/.3166 = 13.4 \text{ years}$$

5. Calculate the total volume of gas produced during the life of the well.

Solution:

Notice that q is expressed in daily units, and a is annual. Either the flow rates must be changed to annual rates, or a must be changed to a daily rate.

$$Q = (q_0 - q_{ec})/a = (700 - 10) \cdot 365/.3166 = 795 \text{ mmcf}$$

6. Calculate the reserves for the well using the shortcut "three years at the current (initial) rate."

Solution:

$$Q = 700 \text{ mcf/day} \cdot 365 \text{ days/year} \cdot 3 \text{ years} = 766.5 \text{ mmcf}$$

11.3.5 Other simulation techniques

Other, more sophisticated reservoir simulations involve such techniques as finite element analyses, finite difference analyses, and other material balance equations which attempt to model the movement of all fluids in the reservoir. Typically, hypothetical water injection and hypothetical production wells are added to the model in an attempt to maximize production from the reservoirs.

In order to do this, the reservoir engineer needs to be able to predict fluid movement in the reservoir rock. This would not be difficult if reservoir rocks were everywhere isotropic and homogeneous. However, no reservoir is homogeneous or isotropic, and fluids always move differently through different parts of the reservoir.

11.3.6 Flow units

In setting up models, the reservoir engineer needs geologic input. The reservoir engineer would particularly like to be able to define units within the reservoir where similar types of flow, called flow units, can be expected. Thus, for large oilfields, the reservoir engineer should be working with the geologist to define flow units that have similar hydraulic characteristics. Overall geometry and internal configuration of reservoirs will be discussed further in Chapters 12 and 13 on depositional systems.

11.4 Examples

11.4.1 Example 1

A geologist comes to you, and wants you to invest $40,000 of your own (or your company's) money in an exploratory well situation. Here is the information that is given to you:

1. Geologic map with bright red arrow showing where the well will be drilled.
2. Cross sections showing the target sandstone.
3. Well logs showing porosity, permeability, and hydrocarbon potential.
4. Your investment is $40,000.
5. For this investment you will get a working interest of 33.3 percent and a net revenue interest of 25 percent.
6. Dry hole cost is shown to be $120,000.
7. Completed well cost is expected to be $180,000.
8. The well is expected to find 120,000 barrels of oil.
9. Probability of success is shown at 20 percent.
10. Price of oil is $25/barrel, no price increase through time and no adjustment for inflation.

The question is whether you should put $40,000 of your hard-earned money in this well. Assuming everything is "on the level," is this a good investment for you or your company, and how do you know?

"On the level" answer Assuming that the situation is legitimate, you should first be aware that the most likely occurrence is that you are going to lose not just some of your investment, but *all* of it. The petroleum industry is a high-risk, high-reward industry and the most likely occurrence is that the well will be a dry hole and you will lose all $40,000. But the prospectus does not say otherwise.

Second, if the well is successful, as a working interest owner you will be asked for 1/3 of the completion cost (1/3 of $60,000 or *another* $20,000) to complete the well before any income can be expected. You may also be liable for 1/3 of any cost overruns (such as blowouts) unless the contract specifically states otherwise.

Third, this one well alone may not produce all 120,000 barrels. Additional wells may be required (let's say three) to produce all of the reserves. It is important that the contract include you in any development wells that result from this discovery. Normally, this is automatic, but if the dealer is unscrupulous, it might not be in the contract. You should also be aware that, as a working interest owner, you will be asked for additional funds to drill additional wells.

Accepting all this, can you still invest in the well? It becomes confusing, because there are a number of possible outcomes. Let's look at the outcomes one at a time:

1. **Dry hole:** If it is clearly a dry hole, you lose $40,000 and take a tax writeoff. If your tax rate is 35 percent, you lose (1 – .35) times $40,000, or $26,000 after taxes.
2. **Complete the well, but the well still won't produce:** You lose $60,000 and take a tax writeoff. If your tax rate is 35 percent, you lose (1 – .35) times $60,000, or $39,000 after taxes.
3. **Successful well:** If it is successful you will need to hire a banker and an accountant, the banker to loan you additional money, and the accountant to keep track of it. It is not that bad. Barring any catastrophic events, the expected income and expenses can be summarized as follows:

	Gross Numbers (100% numbers)	Net to You (income at 25%) (investment at 33.3%)
Expected reserves	120,000 barrels	30,000 barrels (25%)
Expected income at $25/barrel	$3,000,000	$750,000 (25%)
Expected expenses (4 completed wells)	$720,000	$240,000 (33.3%)

In most investments, one can simply compare the investment to the expected profit, and it is fairly easy to determine whether one should invest. However, the petroleum industry has very high risks and additional investment requirements which cloud the picture. One fairly simple way to handle the added investment problem is to look at the risk capacity of the venture. The net risk capacity for this venture is:

$$\text{Risk capacity} = \frac{\text{PVProfit for the whole venture} + \text{DHC}}{\text{DHC}}$$

$$= \frac{(\$750,000 - \$240,000 + \$40,000)}{\$40,000}$$

$$= 13.75$$

Risk capacity compares the profit for the whole venture against the money that is risked. A risk capacity of 13.75 says that if you participate in 13.75 such ventures and one is successful, you will break even. If you believe that the probability of success (Ps) is 0.2 as stated, this should be a good venture because the risk times the risk capacity is considerably greater than one.

Not "on the level" answer What if you don't think the proposer is telling you the truth? There is simply no way to cover all the scams that go on in the petroleum industry. One in particular should be mentioned, however. It is not unheard of for an individual to sell a third for a quarter deal to five, six, or even ten investors. What happens if a promoter sells six third-for-a-quarter deals on one well investment? The promoter collects twice the money necessary to drill the well, and the investors probably don't know the difference unless the well is successful. If the well is successful, then the promoter owes the investors six one-quarter interests or one and a half times whatever comes out of the well. The point is that there are occasions when it is not in the best interest of the promoter to drill a successful well. It is important that you know and trust the individuals that you are dealing with. But, even if you know them and trust them, the most likely occurrence on any one investment still is that you are going to lose all of your money on that investment. On the other hand, one successful well can pay for a number of dry holes.

11.4.2 Example 2

The first example was simple. Now let's consider an example where we believe that the increase in the price of oil will not exactly offset the devaluation of the dollar. Let's take the same example, except that now we'll say that the price of oil is constant, but the value of the dollar is deflating (through inflation) at an effective annual rate of 5 percent.

Figure 11.3 Economic projection using spreadsheet on a personal computer.

(1) Date & Year	(2) Time from First Prod	(3) Prod Rate (BOPD)	(4) Annual Prod (BOPY)	(5) Gross Undis Income (M$)	(6) Net Undis Income (M$)	(7) Time from as of Date	(8) Present Value Discount Factor	(9) Net Discounted Income ($)	Cumulative Net Discounted Profit (M$)
1/1/1995		0.00							
1995			0	($720)	($240)	0.0	1.0000	($240,000)	($240)
1/1/1996	0.0	100.00							
1996			31,534	$788	$197	1.5	0.9277	$182,845	($57)
1/1/1997	1.0	74.08							
1997			23,361	$584	$146	2.5	0.8825	$128,849	$72
1/1/1998	2.0	54.88							
1998			17,306	$433	$108	3.5	0.8395	$90,798	$162
1/1/1999	3.0	40.66							
1999			12,821	$321	$80	4.5	0.7985	$63,985	$226
1/1/2000	4.0	30.12							
2000			9,498	$237	$59	5.5	0.7596	$45,089	$272
1/1/2001	5.0	22.31							
2001			7,036	$176	$44	6.5	0.7225	$31,774	$303
1/1/2002	6.0	16.53							
2002			5,212	$130	$33	7.5	0.6873	$22,391	$326
1/1/2003	7.0	12.25							
2003			3,862	$97	$24	8.5	0.6538	$15,778	$342
1/1/2004	8.0	9.07							
2004			2,861	$72	$18	9.5	0.6219	$11,119	$353
1/1/2005	9.0	6.72							
2005			2,119	$53	$13	10.5	0.5916	$7,835	$360
1/1/2006	10.0	4.98							
2006			1,570	$39	$10	11.5	0.5627	$5,521	$366
1/1/2007	11.0	3.69							
2007			1,163	$29	$7	12.5	0.5353	$3,891	$370
1/1/2008	12.0	2.73							
2008			862	$22	$5	13.5	0.5092	$2,742	$373
1/1/2009	13.0	2.02							
2009			638	$16	$4	14.5	0.4843	$1,932	$375
1/1/2010	14.0	1.50							
2010			$158	$4	$1	15.5	0.4607	$455	$375
	14.2	1.11							
			$120,000	$2,280	$510			$375,005	
		↑ $q_t = q_o e^{-at}$	↑ $Q = (q_n - q_{n+1})/a$	↑ $Q \cdot \$25/bbl$	↑ Times net revenue interest	↑ Times from as of date	↑ e–nj from as of date	↑ Net present value income	↑ Running sum of PV profit to date

This seemingly simple change makes the whole problem considerably more difficult, because now the value of the income from the well is a function of time. Not only must present value adjustments be made for the declining value of the dollar, but now we must know when each barrel of oil will be produced.

Using a spreadsheet and the exponential decline rate equations it is fairly simple to project when each barrel of oil will be produced and make a present value adjustment for that income. Figure 11–3 is such a spreadsheet for the following data:

As of date = 1/1/1995

All investments made on as of date

Production starts on 1/1/1996

Declines at a nominal rate of 30 percent per year

Q = 120,000 bbls gross

q_o = 100 BOPD (4 wells)

q_{ec} = 1 BOPD (4 wells)

Notice now that by making this simple 5 percent adjustment for inflation, the present value profit for the whole venture changes from $510,000 to $375,000, a very significant change. Present value calculations are extremely important in the petroleum industry because the majority of the investment occurs early in the venture, and the income is usually spread out over the next 10 to 50 years.

CHAPTER 12

Clastic Depositional Systems

The subject of depositional systems is a large one, and this book can only cover some of the high points that are relevant to the development geologist. The concept is extremely useful in that it presents a series of unifying models that can be used as predictive tools for understanding overall geometries, internal construction, and lateral extents of major reservoir rock types. It is particularly helpful in understanding the porosity and permeability distributions within major reservoir rock types. This is extremely important for development geologists who are working to develop flow units for reservoir simulation.

This chapter will focus primarily on the geometries, internal configurations, and lateral extents of the various types of common clastic reservoir rocks. It will also help in the identification of aquicludes that occur naturally within many reservoir rock types.

12.1 Depositional Systems

Depositional systems are assemblages of process-related sedimentary environments. Figure 12–1 shows the main clastic depositional systems. This chapter will address each of these depositional systems and their importance to the development geologist. A summary of clastic depositional systems including internal characteristics is given in Table 12–1.

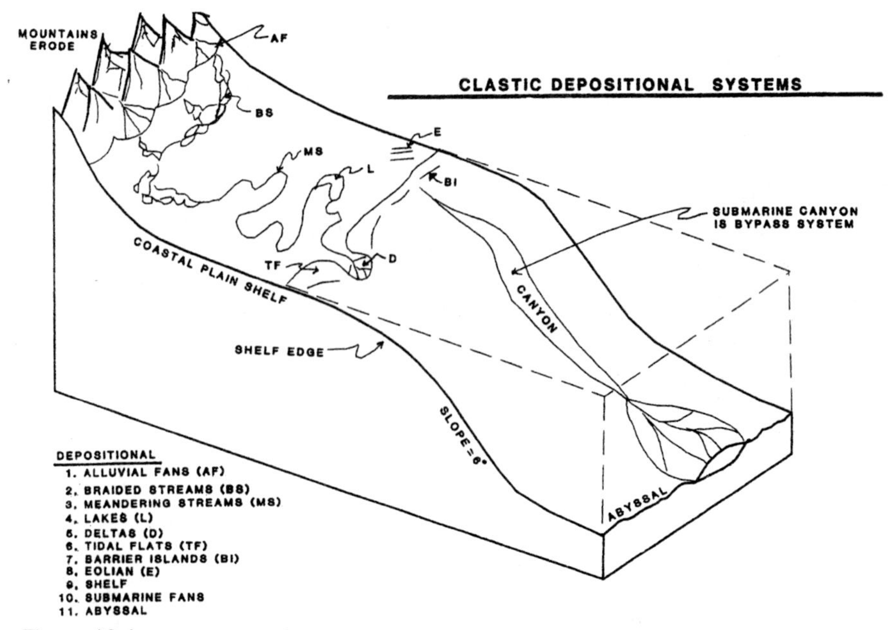

Figure 12.1
Block diagram showing the major clastic depositional systems and their approximate spatial arrangement relative to each other.

12.2 Alluvial Fans

Large alluvial fans do not commonly serve as reservoir rocks for oil and gas throughout the world. They typically occur as large piles of sand and gravel with heterogeneous porosity and permeability. They are deposited in downthrown fault blocks above sea level and are very likely to be altered by diagenesis. If oil or gas is trapped in an alluvial fan, the oil is likely to be degraded by water washing, biodegradation, and/or oxidation depending on subsidence and timing of hydrocarbon migration. Alluvial fans have no natural seals for traps. Shales and evaporites are not typically deposited on an alluvial fan, and seals are rare unless the fan is overlain by an impermeable transgressive unit.

Hydrocarbons can get trapped in relatively small alluvial fans in the form of *granite washes* that occur on the flanks of buried hills or buried mountains. Granite washes occur as transgressive sandstone units that are derived from reworked alluvial material. Typically, granite washes are composed of arkosic conglomerate at the base, relatively pure quartz arenites in the middle, and siltstone and shale or carbonates at

Table 12–1 Summary of sandstone reservoir characteristics in clastic depositional systems.

System	Sequence	Composition	Overall Geometry	Internal Features	Reservoir Quality	Unique
Alluvial fans	Not reliable	Gravel, sand, mud; poorly sorted	Big pile of gravel; fine outward	Crude bedding	High vertical and horizontal phi & K	Shale is rare; no seals for traps
Braided streams	Blocky or fine upwards	Gravel and/or sand; little shale	Braid belt parallels depositional dip sheets on pediment surf	Internal pods	Outstanding	Not particularly common; exception is Prudhoe Bay Sadlerochit Formation
Meandering streams	Fine upwards	Mud system with point bar sands	Meander belts in organic rich muds	Upper part fine grained and shingled	Base is excellent; top has poor vertical and lateral permeability	Clay plugs; shingling by clay drapes
Eolian	Not reliable	Sand and silt	Not reliable	Tabular and festoon cross beds	Excellent but not uniform	Difficult to water flood because of channeling
Gilbert delta	Coarsens upwards	Sand, gravel, and mud	Dependent on depositional setting	Topset, foreset, bottomset	Excellent	Rare geologically
Elongate delta front	Coarsens upwards	Sandstone	Elongate parallel to depositional dip	Distributary mouth bar	Very good	Likely to be overprinted by point bar sequence
Lobate delta front	Coarsens upwards	Sandstone	Sheet sand	Distributary mouth bar, beach, and shoreface	Excellent	At least ¼ of the world's oil and gas
Distributary channels	Fine upwards	Sandstone	Linear, unpredictable	Override delta front	Excellent	May appear as fine upwards in coarsens upwards with different water level
Crevasse splays	Thin, coarsen upward in interdistributary	Sandstone, siltstone	Thin, local sheets	Veining in web of bird foot	Locally good	
Cuspate delta and barrier island	Coarsens upwards	Sandstone	Long, linear parallels depositional strike	Fossils and burros, eolian likely on top	Outstanding	Encased in source rocks
Proximal submarine fan	Overall coarsen upward; individual beds fine upwards	Sandstone and occasional gravel	Pod shaped	Pods may not be connected	Excellent	Likely to be overpressured
Distal submarine fan	Bouma turbidites—fine upwards	Sandstone, fine to very fine grained	Overall pod shaped	Very thin turbidites	Individually poor; thick sequences OK	Each sand is its own reservoir and its own water level

the top. The quartz arenites and conglomerates make good reservoir rocks that commonly grade upward into poor reservoir rock such as siltstone or shale. In other cases the sandstone units may grade upward or laterally into carbonate units, such as in the Panhandle portion of the Panhandle-Hugoton field in North Texas (Pippin, 1970).

12.3 Braided Streams

To most people the term *braided stream* brings a mental image of the Platte River in Colorado or Nebraska or the Arkansas River in Oklahoma (Fig. 12–2). While these are true braided streams, it is difficult to imagine them as forming major, blanket reservoir rocks because they are relatively small, they must fit within the confines of a river valley, and they are primarily pass-through systems for sediments. They are not major depositional systems. Given this mental picture, how can braided streams be important in forming widespread sheet deposits of sand and gravel that are among the most significant reservoir rocks in the world?

Although the Platte and Arkansas Rivers are braided streams, they are not the correct model for major oil and gas reservoir rocks. Mantled pediment surfaces form some of the most significant reservoir rocks in the world, including the Sadlerochit Formation of the Prudhoe Bay field, the largest oilfield in North America.

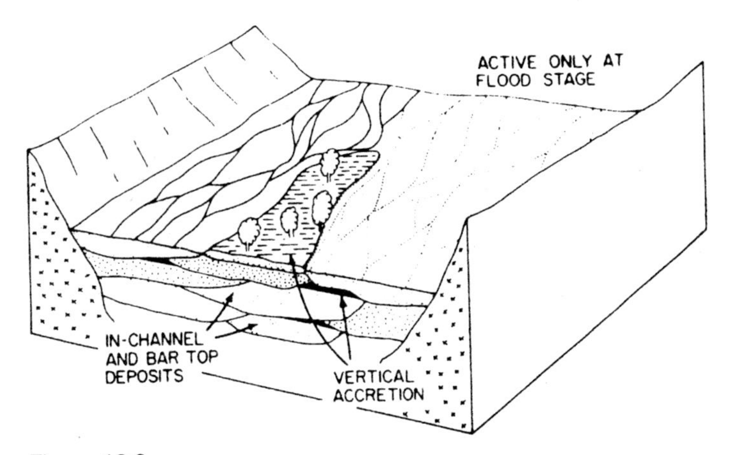

Figure 12.2
Block diagram of a braided stream sequence. From Walker & Cant (1984), reprinted by permission of the Geological Association of Canada.

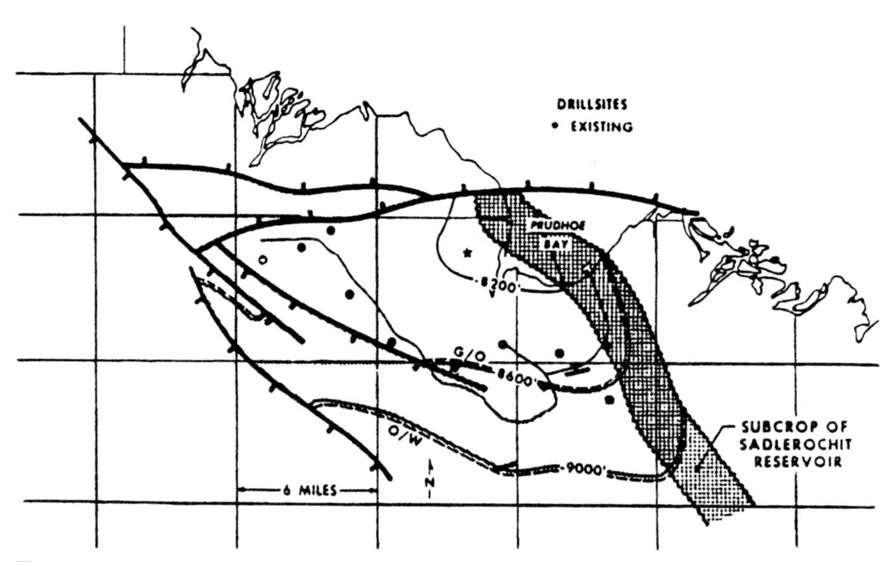

Figure 12.3
Structure contour map on the top of the Sadlerochit Formation of the Prudhoe Bay field showing the subcrop area for the formation. From Morgridge and Smith (1972), reprinted by permission of AAPG.

Braided stream sequences called *mantled pediments* form sheet sand and gravel deposits that slope away from some major mountain ranges around the world. Pediment surfaces are both erosional and depositional. That is, during major floods, currents at the base of braided stream channels are able to erode the underlying sediments such that the base of a mantled pediment is always erosional. As channels relocate and as floods wane, sand and gravel are deposited across a broad plain that slopes away from the major mountain range. If the sequence becomes buried in the subsurface, it has the potential to become a super reservoir rock that is bounded both above and below by unconformities.

The Sadlarochit Formation of the Prudhoe Bay field is a delta and braided stream sequence that is bounded by unconformities and was derived from a Permian and Triassic mountain range that existed to the north of the field (present Arctic Ocean). The trap is both structural and stratigraphic (Fig. 12–3), and the reservoir rock is composed of approximately 600 feet of almost pure sandstone (Fig. 12–4). Porosities for the field range between 20 and 24 percent. Permeabilities range from 300 millidarcies to 2 darcies.

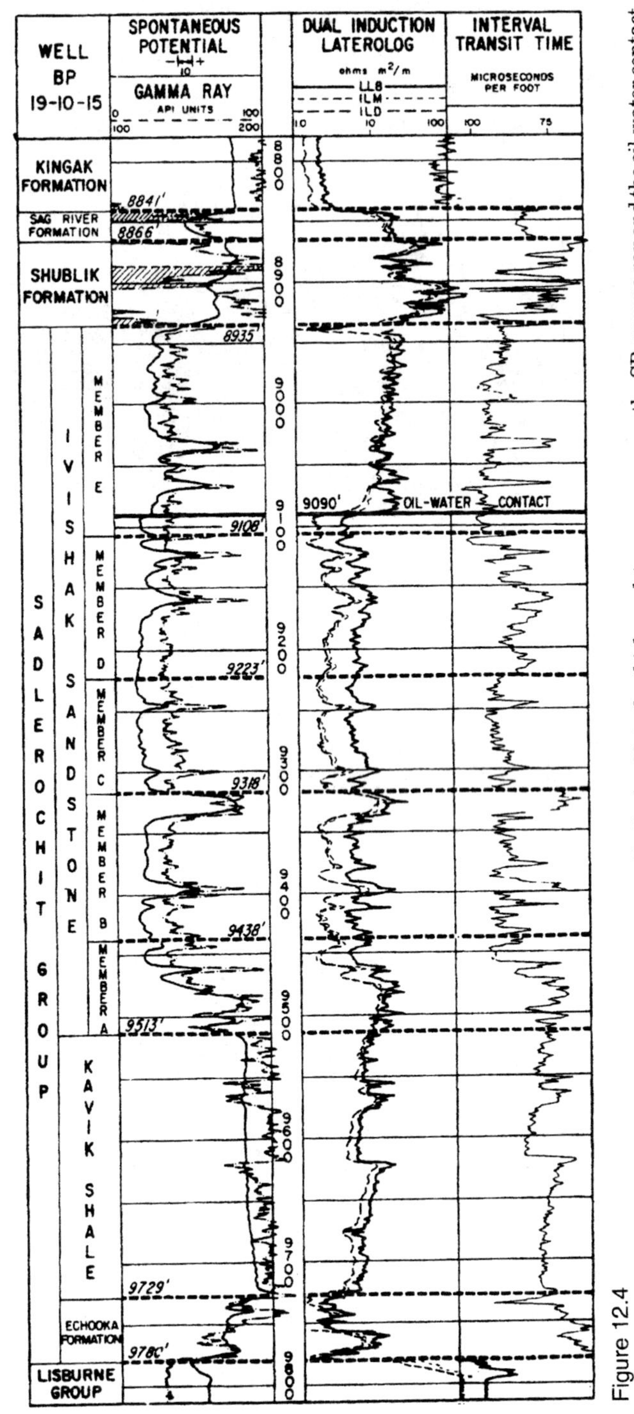

Figure 12.4
Electric log of the Sadlerochit Formation, Prudhoe Bay Field, Alaska. Note the thick sandstone sequence on the SP-gamma ray and the oil-water contact on the resistivity curve. From Jones & Speers (1976), reprinted by permission of AAPG.

12.4 Meandering Streams

Meandering streams are primarily mud sequences with embedded point bar (sandstone) sequences (Figs. 12–5 and 12–6). Point bar sequences are coarsening upward sandstone units that form by lateral accretion, with deposition occurring on the inside of bends of meander loops. As the river floods, erosion occurs on the cutbank (outside) of the meander loop, deposition occurs on the inside (point bar side), and then the flood abates. The base of the sequence is erosional, commonly with gravel or coarse sand above the erosional surface. As the flood abates, the velocity of the stream slows significantly and *clay drapes* are deposited such that the upper part of the point bar sequence may be shingled or partitioned into separate stratigraphic traps by these thin clay layers.

In the Pennsylvanian of Missouri and Kansas, such partitioning can result in very unusual hydrocarbon distributions. Because the clay drapes and enclosed reservoirs dip at a low angle, a vertical well may encounter a wet sand above hydrocarbons separated by a clay drape that may be no more than a couple of inches thick.

A second important point is that when meander loops are abandoned, two results can occur.

1. If the loop is abandoned slowly, it fills with sand and silt as velocities in the channel slowly subside.

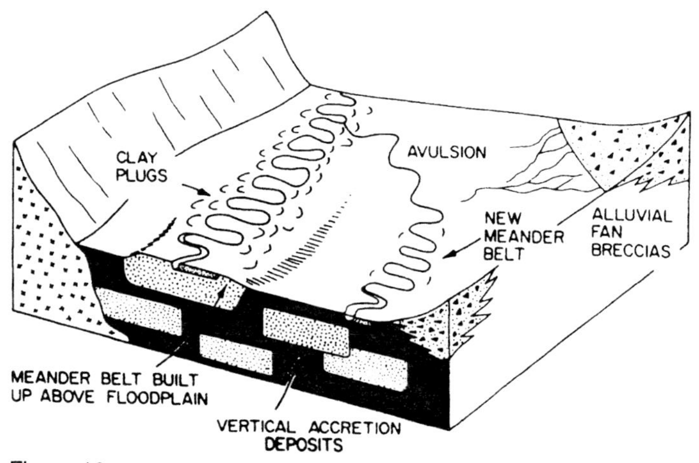

Figure 12.5
Block diagram of a meandering stream sequence. From Walker & Cant (1984), reprinted by permission of the Geological Association of Canada.

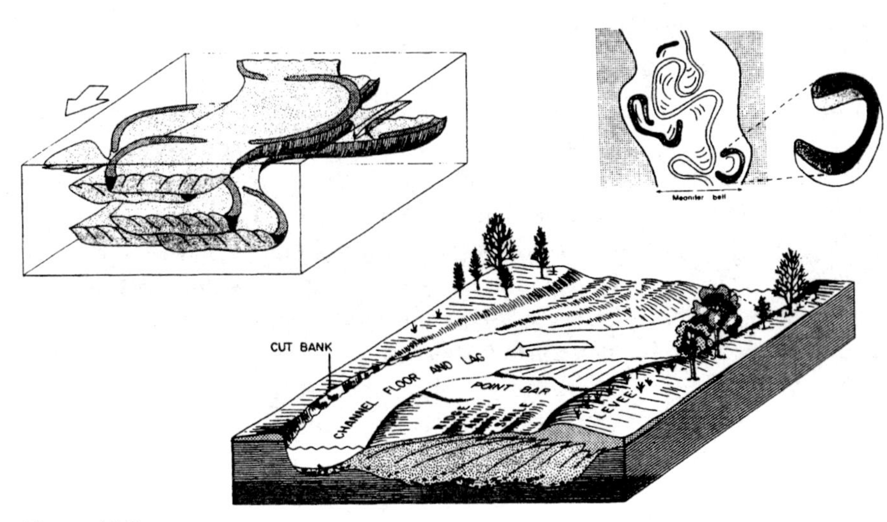

Figure 12.6
Block diagrams of meandering stream sequences. From Galloway & Hobday (1983), reprinted by permission of Springer-Verlag.

2. If the loop is abandoned quickly, it becomes part of the flood plain. It fills slowly with mud during river flood stages and with organic material during nonflood stages. When this happens, a horseshoe-shaped, impermeable *clay plug* forms that naturally surrounds an excellent reservoir rock on three sides. If the fourth side becomes closed off, or the rocks are tilted slightly, an excellent stratigraphic trap is formed.

Such traps are common in the Pennsylvanian of north Texas, eastern Oklahoma, and eastern Kansas. It is a very difficult trap to predict from the surface, but can occasionally be identified by three dimensional seismic.

12.5 Lakes

Lakes are considerably more complicated and diverse than most people realize. Deep lakes, at some time in their history, almost always develop a *thermocline*—a depth below which a temperature discontinuity occurs. Above the thermocline the water is warm, agitated, and oxygenated. Below, the water is cold, relatively still, and anoxic, ideal for preservation of organic material. In shallow lakes a thermocline may never be established, or if established, it may only be temporary as storms may destroy it. In deep lakes, thermoclines may be permanent or temporary depending on whether the lake freezes.

Before a lake can freeze, water in the lake must turn over because fresh water is most dense at 4°C, just before it freezes. As the water turns over, the thermocline is temporarily disrupted, and many lakes turn over several times before freezing. However, if a deep lake does not freeze, the thermocline is normally permanent. Such lakes commonly occur in temperate climate where algal blooms are also likely. As the algal material dies and settles through the thermocline, it is preserved in the oxygen-poor sediments below the thermocline. Such lakes such as the Green River oilshales of Colorado form some of the best oil source rocks in the world. Highly saline lakes teem with biologic activity and also make good source rocks for oil.

Reservoir rocks are a different story. With the exception of some lake deposits in China, lake deposits are not noted for the formation of good reservoir-quality rocks. Gilbert or fan deltas are found on the edges of some lake deposits, but they are not commonly associated with good seals.

Altamont-Bluebell (Lucas & Drexler, 1986), on the south flank of the Uinta Mountains, occurs in the Wasatch Formation, which is lacustrine, but it is quite unusual in that much of the reservoir rock is fractured shale, siltstone, and some sandstone (Fig. 12–7). The shale is considered to be an oil wet source rock and the main production zone is highly overpressured relative to sediments both above and below. Altamont has a matrix porosity ranging between 3 and 7 percent, with an average matrix permeability of .01 millidarcies. Yet the natural fracture system allowed wells to flow initially at rates of up to 1,000 BOPD. Optimum well spacing for the field is 640 acres, which is very unusual for oil production.

12.6 Deltas

It has been estimated that approximately 40 percent of the world's oil and gas production comes from deltaic rocks. Deltas can be classified as follows:

1. Fan deltas. Fan deltas occur where alluvial fans form immediately adjacent to the ocean and in lakes where fresh water enters fresh water. In cross section most fan deltas look like Gilbert deltas with topset, foreset, and bottomset beds (Fig. 12–8).

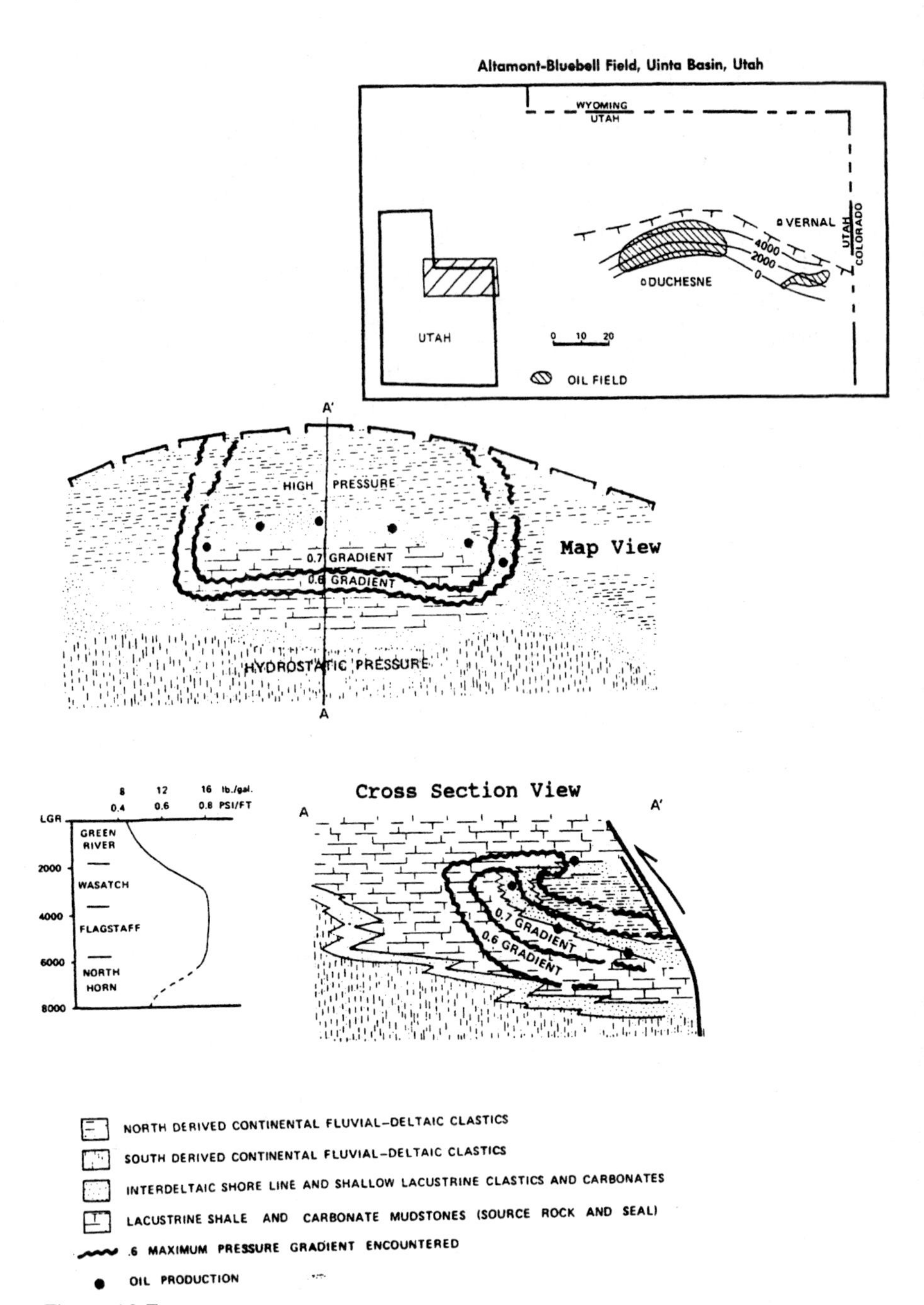

Figure 12.7
Map view and cross section through the Altamount-Bluebell Field, Utah, showing the unusual pressure distribution for the field. From Lucas & Drexler (1976), reprinted by permission of AAPG.

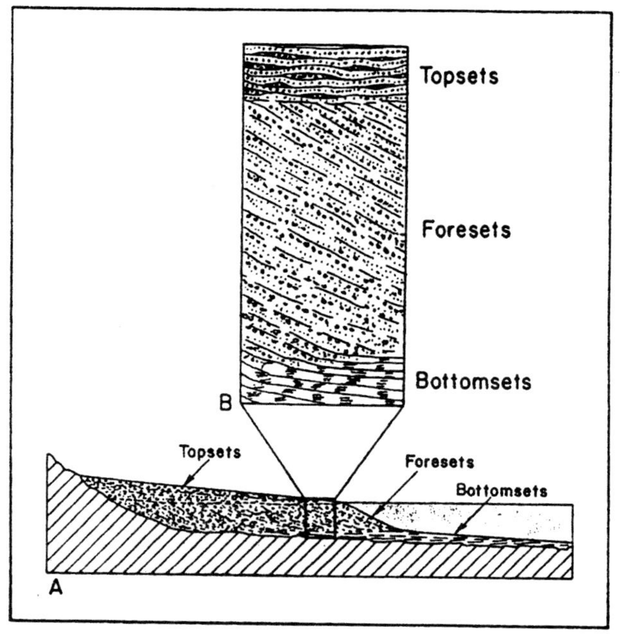

Figure 12.8
Cross section through a Gilbert delta. From Friedman, G.M., and Sanders, E., Principles of Sedimentology, p. 502. Copyright 1978 John Wiley & Sons, reprinted by permission.

2. High-constructive elongate deltas (Figs. 12–9 and 12–10).
3. High-constructive lobate deltas (Figs. 12–9 and 12–10).
4. Wave-dominated delta (high-destructive in Figs. 12–9 and 12–10).
5. Tide-dominated deltas (Fig. 12–10).

Fresh water, even when laden with mud, is considerably less dense than sea water. When fresh water enters sea water at a delta, the fresh water floats on top of sea water and spreads out (hypopycnal flow). Silt and sand are deposited at the mouth of the river in a distributary mouth bar, but clay particles stay in suspension and float in the fresh water on top of sea water out to sea. As waves mix the fresh water with sea water, the electrolytes in sea water cause the clay particles to flocculate (clump together) and sink to the bottom as silt-sized particles.

The important point is that silt and sand become separated from clay particles at the mouth of the river. Sand and silt at the mouth of the river tend to be reworked by waves, and are moved along the shoreline

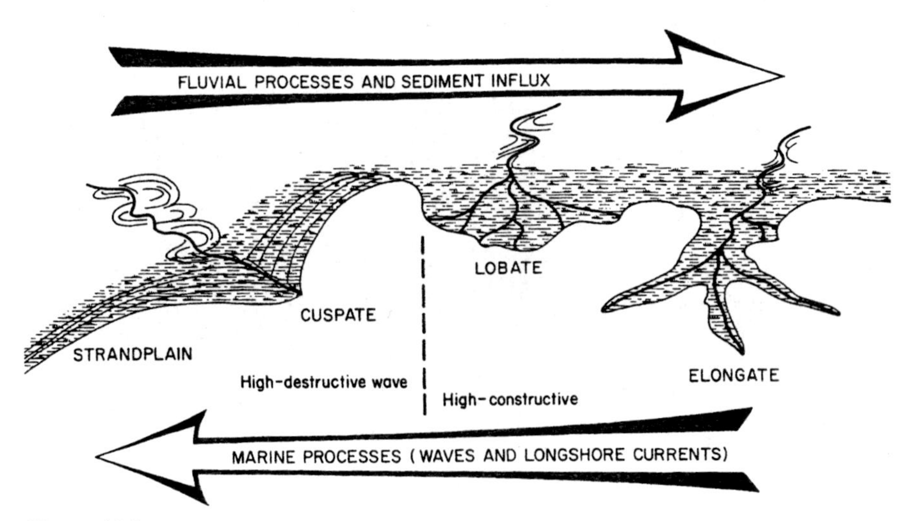

Figure 12.9
Morphologies of three main marine delta types as a function of fluvial sediment influx relative to waves and longshore processes. From Fisher et al. (1969).

by longshore currents, while clay-sized particles are carried out to sea. Because most rivers carry a lot of mud, one of the most recognizable deltaic facies is the prodelta mudstone. Virtually all major deltas have a large pile of mud, clay, or shale in the associated marine environment. These clays grade laterally into reservoir-quality rocks in various and complex relationships depending on the delta type.

Delta types also determine the geometry of the principal reservoir rock types. The following sections describe the overall geometry and internal configuration of different sand types associated with different delta types.

12.6.1 High-constructive elongate delta

A high-constructive elongate delta is a river-dominated delta that typically occurs in a shallow, slowly subsiding basin. It is a delta that typically builds seaward very rapidly across a shallow basin. The deltas associated with most Pennsylvanian cyclothems of north Texas, Illinois, Missouri, Kansas, and Oklahoma are mainly high-constructive elongate.

Such deltas commonly have a relatively narrow delta front sandstone (Fig. 12–11), on the order of 2 to as much as 20 miles across. They advance seaward very rapidly and leave large areas of interdeltaic lagoons, estuaries, and tidal flats between the deltas. As a prograda-

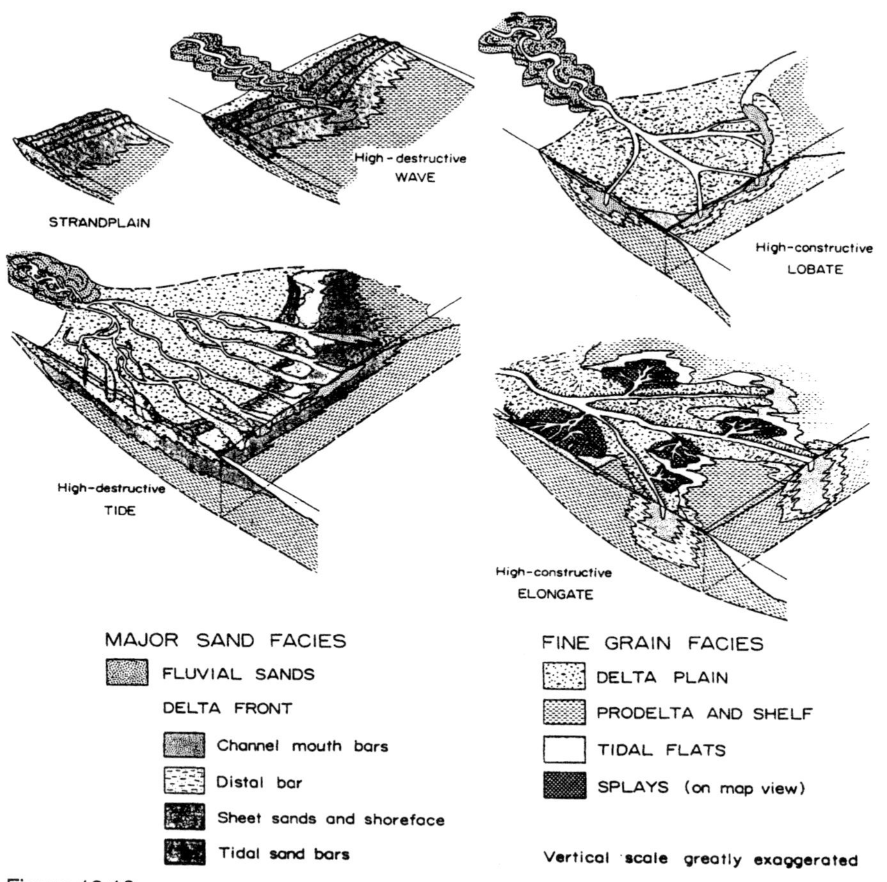

Figure 12.10
Morphologies of four main marine delta types. From Fisher, et al. (1969), reprinted by permission.

tional sequence, the delta front sandstone is usually a coarsening upward (CU) sequence (Fig. 12–12) that is commonly reworked by meandering stream sequences (fining upward sequences, with disconformable bases) that prograde over them.

Meander belts tend to stay within the confines of the original delta front sandstones because meandering channels are able to erode delta front sand more easily than interdistributary mud. On this type of delta, then, the delta front sandstones are not often found because they have been replaced by point bar sequences (Fig. 12–13).

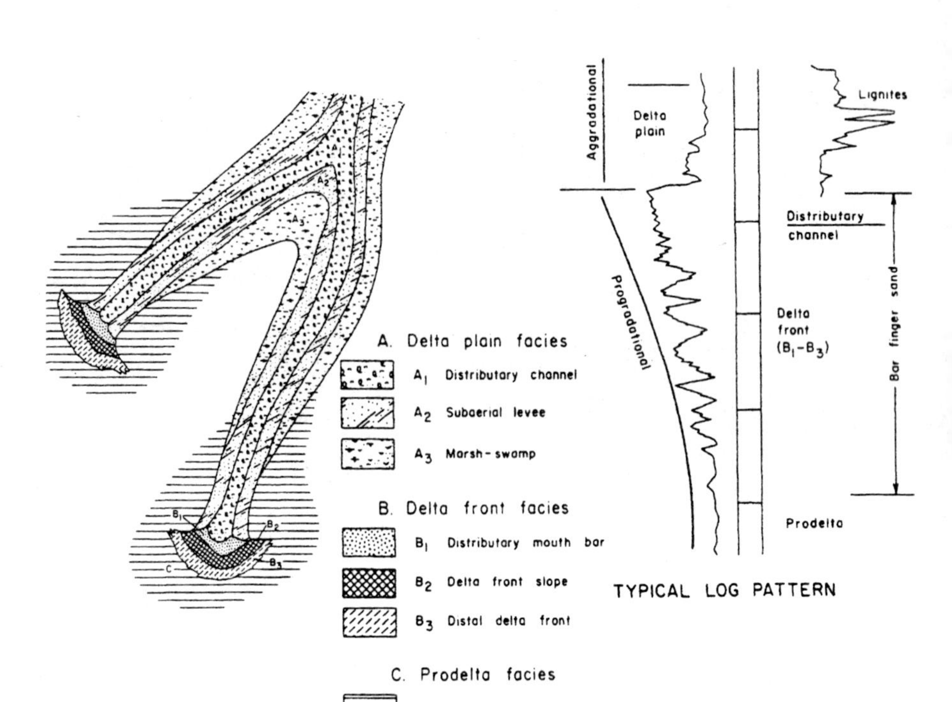

Figure 12.11
Map view showing the morphology of a high-constructive elongate (river-dominated) delta. From Fisher et al. (1969), reprinted by permission.

12.6.2 High-constructive lobate delta

The high-constructive lobate delta has characteristics of both the river-dominated and the wave-dominated delta. It is a hybrid. It is probably the most common delta type in the geologic record, and most deltas of the Texas and Louisiana Gulf Coast are of this type. Figure 12–14 shows the facies that are commonly found on this type of delta. Notice that the distributary channels are straight and that the distributary mouth bar has now been spread out by wave and longshore drift action into a long sand body, on the order of 25 to 50 miles long. As the delta progrades, the delta front sandstone will form a sheet sandstone that can be correlated over large distances. These are the correlatable sandstone units in the Texas and Louisiana Gulf Coast. Following the progradation of the delta front sandstone is the delta plain with distributary channel sandstones and

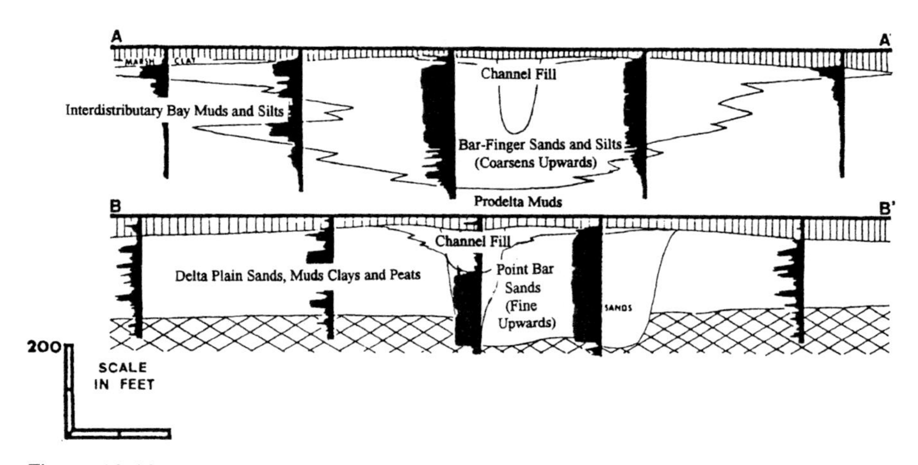

Figure 12.12
Representative electric-log cross sections through the Holly Springs delta system, Lower Wilcox of Louisiana, showing typical electric-log responses for different parts of the delta. From Fisher et al. (1969), reprinted by permission.

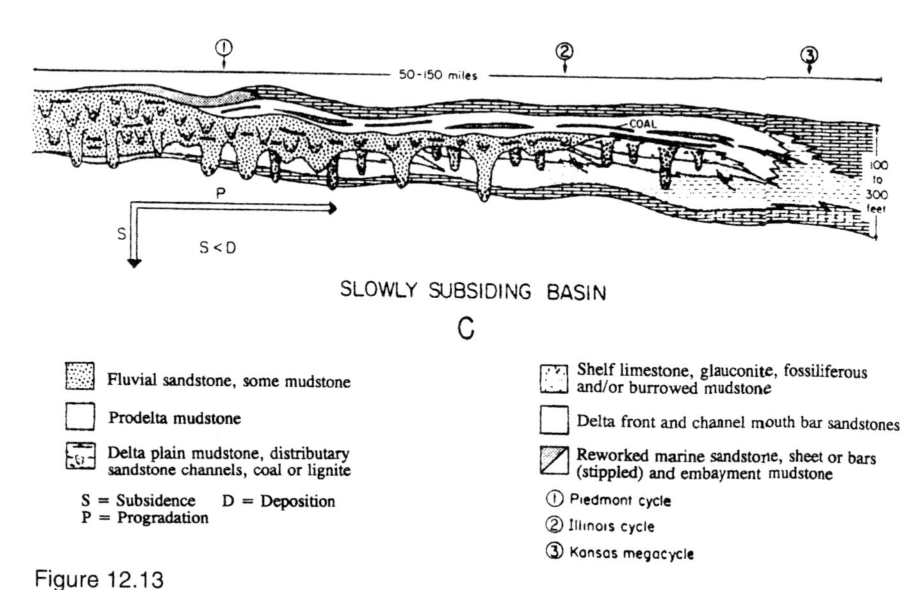

Figure 12.13
Cross section through a Pennsylvanian cyclotherm showing how the delta front sandstone is commonly overprinted by point bar sequences. From Brown (1980), reprinted by permission of AAPG.

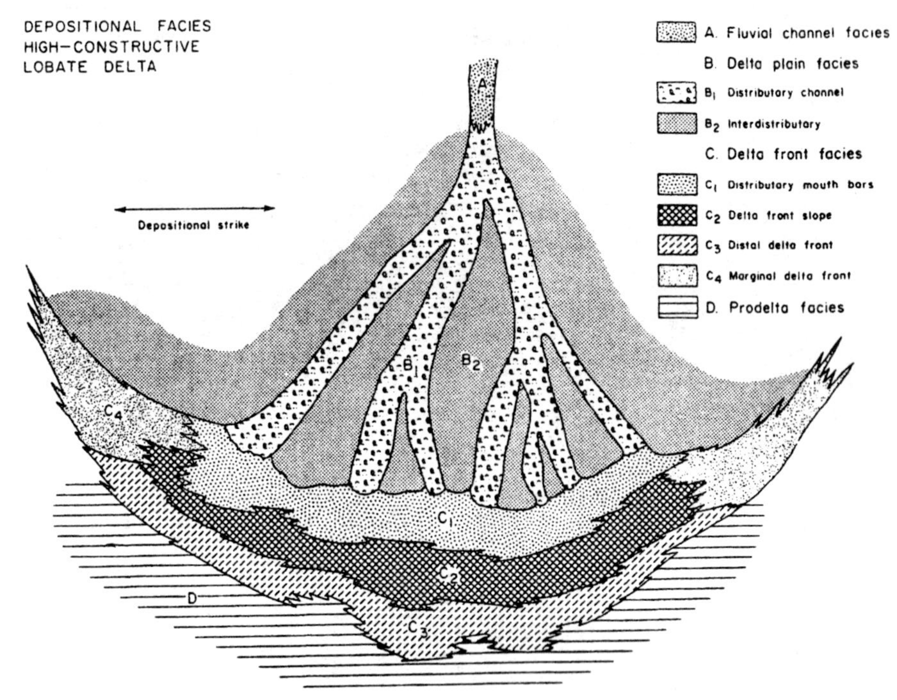

Figure 12.14
Map view of a high-constructive lobate delta showing the principal facies. From Fisher et al. (1969), reprinted by permission.

interdistributary organic rich marshes, lakes, swamps, lagoons, bayous, and other environments that will normally contain some coal.

The delta front sandstone will tend to be a coarsening upward (CU) sandstone, and the distributary channels will tend to be fining upward (FU), as shown in Figure 12–15.

12.6.3 Wave-dominated deltas

Wave-dominated deltas look very much like barrier island sequences. They prograde seaward through time, but, because they are not dominated by major rivers, there is not usually enough sediment influx for the delta front sandstone to form major blanket sandstone. Instead the sandstone usually merges with a barrier island sequence to form a long, linear sandstone body parallel to the shore line. Most of the Cretaceous deltas of the western interior seaway such as the Muddy and Frontier Formations were probably formed by wave-dominated deltas.

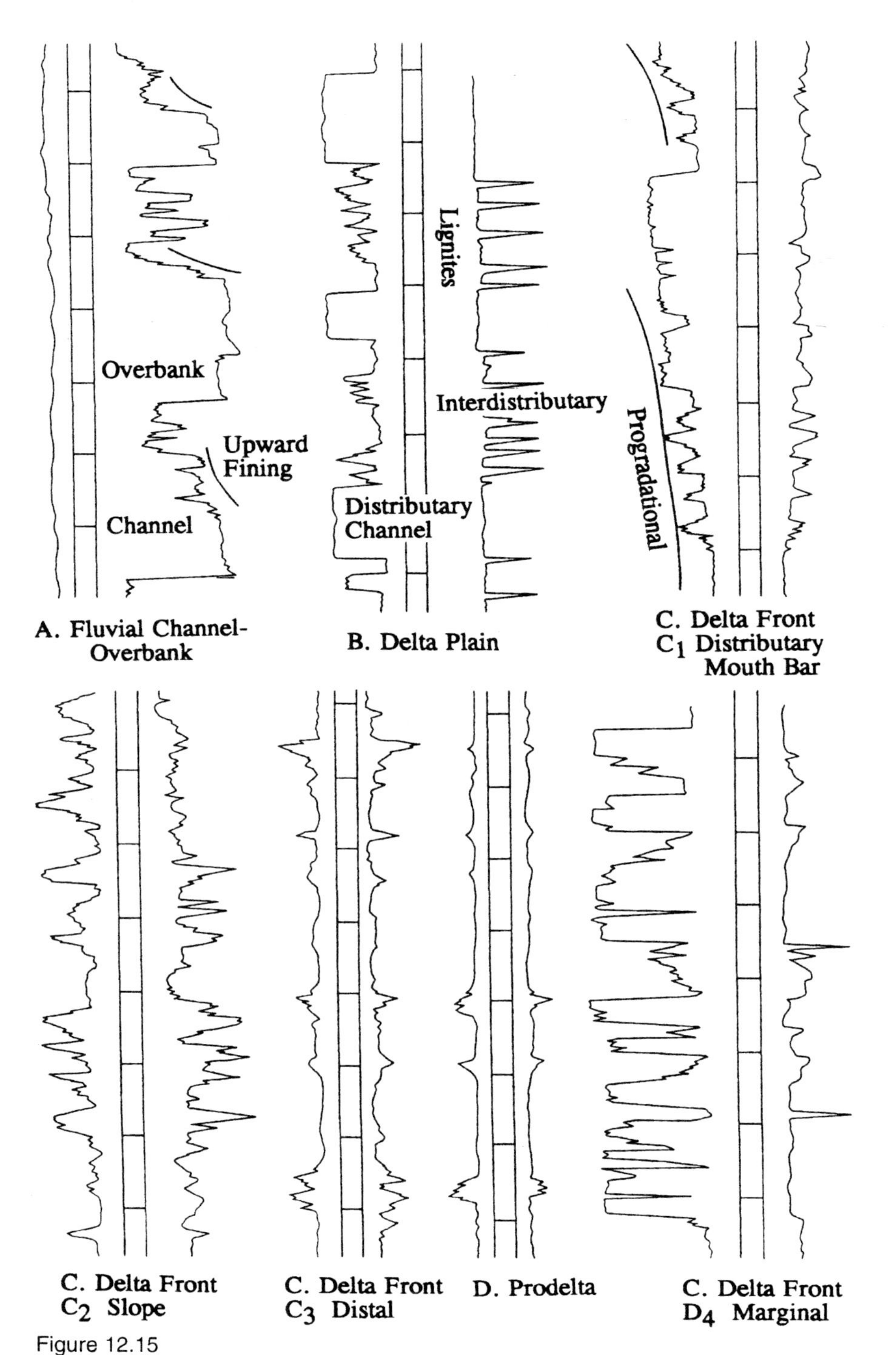

Figure 12.15

Typical electric log responses to different deltaic environments. From Fisher et al. (1969), reprinted by permission.

12.6.4 Tide-dominated deltas

Many of the largest deltas of the modern world, such as the Mekong, Brahmaputra, and Ganges, have strong tidal influences. The delta front sandstone on such deltas is modified significantly by tidal currents, such that it is typically a submarine sandstone that shows strong cross bedding both towards the sea and away from the sea.

Although many of the largest deltas of the modern world are tide-dominated, there seems to be little documentation for ancient tide-dominated deltas in the geologic literature. Tide-dominated deltas commonly occur in drowned river valleys associated with large estuaries where they can be easily destroyed by changes in sea level. If sea level drops, they are destroyed by subaerial erosion; if sea level rises, they are eroded by submarine canyons. They are not commonly preserved in the geologic record.

12.7 Transgressive Sandstones

Most major deltas occur in areas that are or were subsiding. Delta lobes tend to build seaward, become abandoned, and founder in the subsiding basin. During the subsidence phase, waves attack the front of the delta, and commonly a sandstone derived from wave destruction transgresses across the top of the abandoned delta.

Transgressive sandstones occur in areas where relatively little clastic influx occurs—in a sort of blue-water environment where organisms thrive. Transgressive sandstones commonly contain numerous calcareous fossils such as oysters and gastropods. During diagenesis the carbonate shell fragments are commonly partially dissolved and reprecipitated in pore spaces, and the resulting sandstone is commonly a carbonate-cemented sandstone. Such a sandstone will give “thumbs out” on the electric log and should not be mistaken for hydrocarbons. Porosity logs will show it to be tight. When looking at older fields where there are no porosity logs, however, watch out for the thin (usually less than five feet thick) resistive calcareous sandstone that occur at the top of many delta front sandstone units. It is nothing more than the conclusion to a normal deltaic progradation-transgressive cycle.

12.8 Transgressive Valley Fills

A great deal of oil has been produced from a number of different sandstones from the Pennslyvanian cyclothems across Kansas and Oklahoma. Such sandstones as the Bartlesville, Warner, Bluejacket, Prue, and Squirrel Sandstones are channel-like sandstones that are well known for being discontinuous laterally and very difficult to correlate. The term Squirrel comes from the fact that the sandstones are "squirrely" when it comes to correlations. These sandstones have been variously interpreted as channel fills, braided stream sequences, and point bar sequences that have overridden delta front sandstones.

Recent work suggests that many of these sandstones may be valley fill sequences. During low sea level stands, low relief stream valleys developed across the exposed land surface. As sea level rose, the valleys appear to have become filled with distributary mouth bars, delta front sandstones, and sand derived from wave action along the shoreline. The sandstones have very little cross bedding and good lateral permeability. Valley fill sequences are probably more important than has previously been recognized in the geologic literature.

12.9 Tidal Flats

Flanking most major delta systems of the world are areas of major mud deposition that have variously been called tidal flats, estuaries, chenier plains, and strand plains. These areas may contain thin chenier ridge sandstones or small concentrations of sand in flaser beds or tidal channels. For the most part, these areas are not conducive to the formation of a broad, easily found, extensive reservoir sandstone. This environment is composed primarily of mud and may act as a good seal or good source rock.

12.10 Barrier Islands

Laterally adjacent to and grading into the strand plain system is the barrier island system. Figure 12–16 shows a cross section through the Galveston Island barrier island system, and Figure 12–17 shows typical cross sections through transgressive, regressive, and aggradational barrier island systems. In the geologic record, by far the most important model is that of the regressive barrier island similar to Galveston Island and Padre Island.

Galveston Island and Padre Island are barrier island complexes that build seaward or prograde through time. They simply do not have enough sand input to prograde over large distances like a delta front sand. Therefore, the barrier island complex preserved in the geologic record is a long, linear sand body (not a sheet sand), formed parallel to the shoreline, and sandwiched between marine and lagoonal, organic

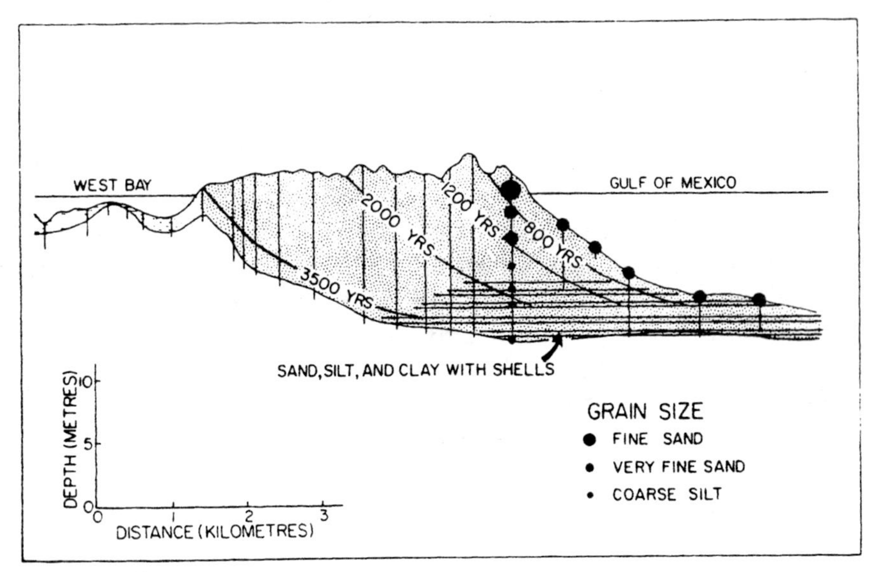

Figure 12.16
Cross section through the prograding Galveston barrier island. From Bernard et al. (1962), reprinted by permission of Geological Society of America.

rich mudstone. It is likely to be an excellent reservoir rock, several miles across, that is sandwiched between excellent and relatively impermeable source rocks. It also forms its own natural stratigraphic trap. Although barrier island sequences may be difficult to locate initially, once discovered, they are truly excellent trends to follow for production. Examples include the parts of the Frio and Wilcox trends of Texas and Louisiana, the Pennsylvanian shoestring sands of Kansas, and parts of the Cretaceous interior seaway (Fig. 12–18).

12.11 Sandstone Geometries

As a river enters the ocean, sand becomes separated from mud. Mud is carried out to sea and is deposited on the shelf or the continental slope. Sand is deposited in distributary mouth bars, and is reworked by storms, waves, and longshore drift. Sand is deposited on delta fronts in overall geometric patterns similar to those shown in Figure 12–19, and may then be carried along the shoreline and deposited in patterns similar to those shown in Figure 12–20.

In the geologic record there are many sandstone bodies that have been called *shelf sandstones,* particularly in association with carbonate rocks. Extensive shelf sandstones are difficult to explain because there seem to be relatively few modern analogies. It is known that storms attack barrier island sequences and delta front sandstones. Turbidity currents associated with storms are capable of carrying sand as much as 5 miles offshore on the shelf, but not as far as 50 miles (except through submarine canyons to deep water environments). Swift (1973) has shown

A. TRANSGRESSIVE

MARSH
BEACH
WASHOVER
LAGOON
MARSH

B. REGRESSIVE

BEACH
LAGOON
SHOREFACE
OFFSHORE

C. AGGRADATIONAL

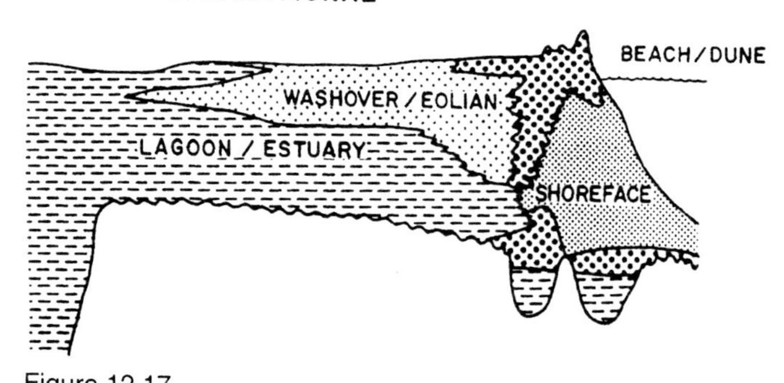

Figure 12.17
Cross sections through transgressive, regressive, and aggradational barrier island systems. From Galloway & Hobday (1983), reprinted by permission of Springer-Verlag.

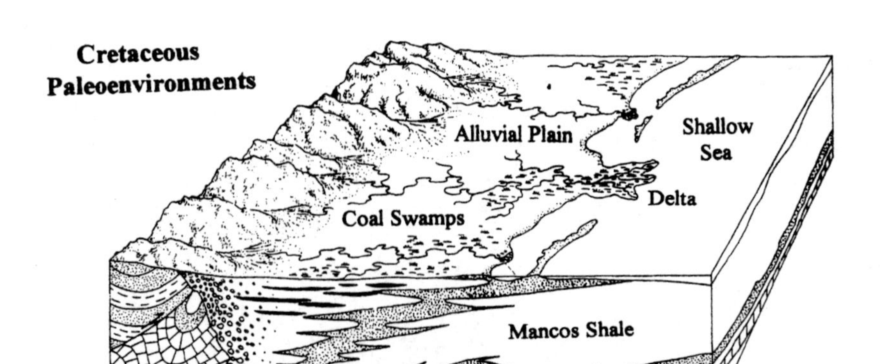

Figure 12.18
Block diagram of the western side of the Cretaceous interior seaway. From Hintze (1982).

a number of different types of sandstones that occur on the shelf off the east coast of the U. S. (Fig. 12–21). Many of these sand bodies are barrier islands or transgressive sandstones that were formed during low sea level stands and have now simply been drowned by higher sea level stands.

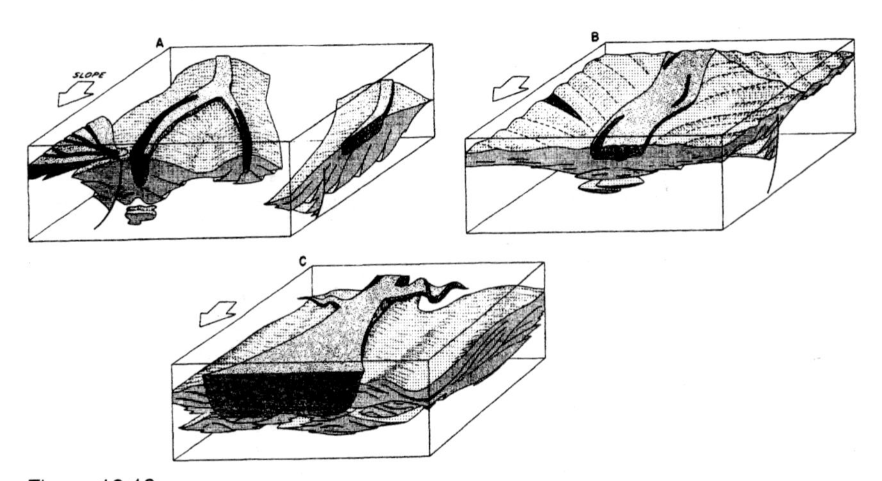

Figure 12.19
Schematic three-dimensional sand geometrics for (A) river-dominated delta system; (B) wave-dominated delta; (c) tide-dominated delta. From Galloway & Hobday (1983), reprinted by permission of Springer-Verlag.

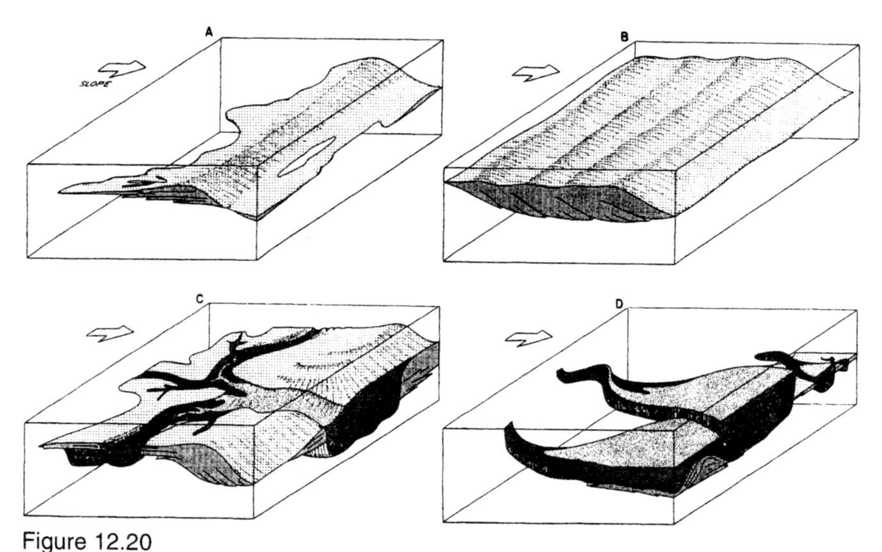

Figure 12.20
Schematic three-dimensional sand geometrics for (A) trangressive barrier island; (B) regressive barrier island; (C) barrier bar and inlet complex; (D) estuary and subtidal sandstone complex. From Galloway & Hobday (1983), reprinted by permission of Springer-Verlag.

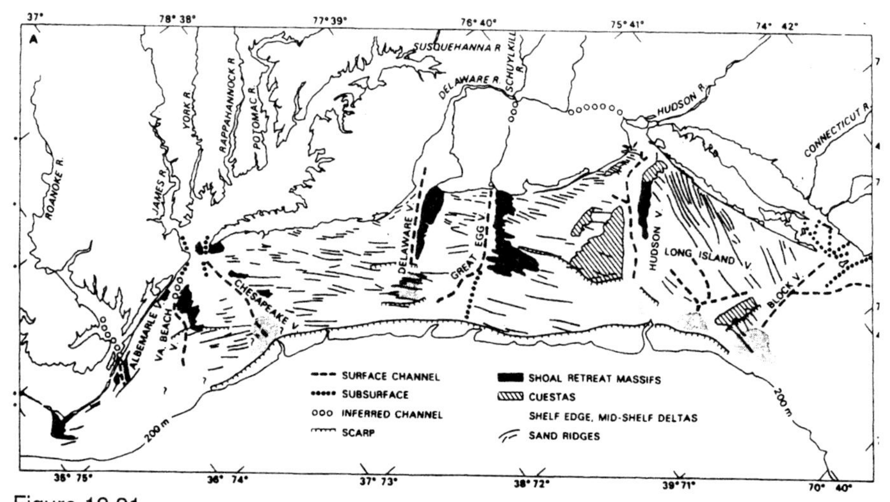

Figure 12.21
Major morphologic features of the Middle Atlantic Bight. From Swift et al. (1973).

12.12 Submarine Fans

Submarine fans occur in deep ocean basins. They are fed by clastic material held in autosuspension by turbidity currents that flow primarily down submarine canyons (Figs. 12–21 and 12–22). Although turbidity currents can occur anywhere in a lake or in the ocean, the largest fans all occur in association with major submarine canyons that feed the fans.

The proximal part of the fan is composed primarily of sand-sized material that was transported down the canyon by slow creep and autosuspension. Proximal sandstones commonly show soft sediment deformation, as the sand-sized material was often deposited originally in a loose

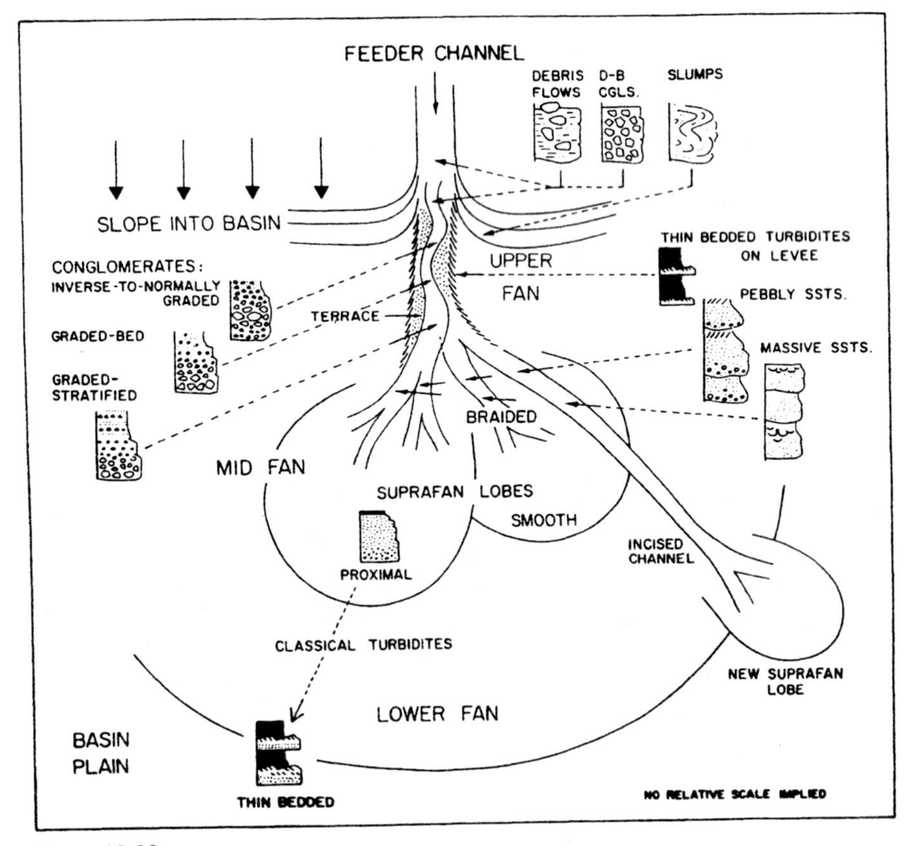

Figure 12.22
Diagram showing major facies that occur in a submarine fan. From Walker (1979), reprinted by permission of the Geological Association of Canada.

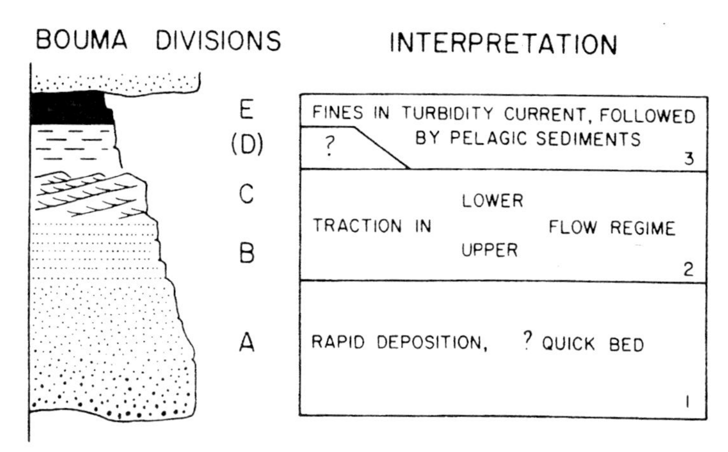

Figure 12.23
Five subdivisions of the ideal Bouma turbidite. From Walker (1984), reprinted by permission of the Geological Association of Canada.

packing arrangement. During dewatering the loosely packed sand became rearranged by small-scale slumping and grain flows.

The distal part of the fan is composed of alternating thin sandstones and shales that have formed from the flushing of major turbidity currents through the system. Such currents flow through major distributary channels (including levees) down to a distal sublobe of the fan where currents spread out, slow down, and deposit their loads. If the full spectrum of currents is recorded, the full Bouma turbidite (Fig. 12–23) is deposited. More commonly, only parts of the sequence are deposited, and the principal characteristic of a submarine fan sequence is thousands of feet of alternating thinly bedded (inches to feet thick) sandstones and shales. Through geologic time, most shelves prograde seaward, and the result is typically an overall coarsening upward sequence as shown in Figure 12–24, although each individual turbidite is a fining upward bed.

12.12.1 Forties Field

Many North Sea fields have reservoirs in submarine fans. The Forties Field (Figs. 12–25 and 12–26) is an example of a field developed in the proximal to middle part of the fan. The Main sand unit is approximately 150 feet thick and shows a very blocky log signature (Fig. 12–26, top). Notice that this unit has a consistent water level, and a water drive, and its volumetrics would be handled like any other sandstone reservoir.

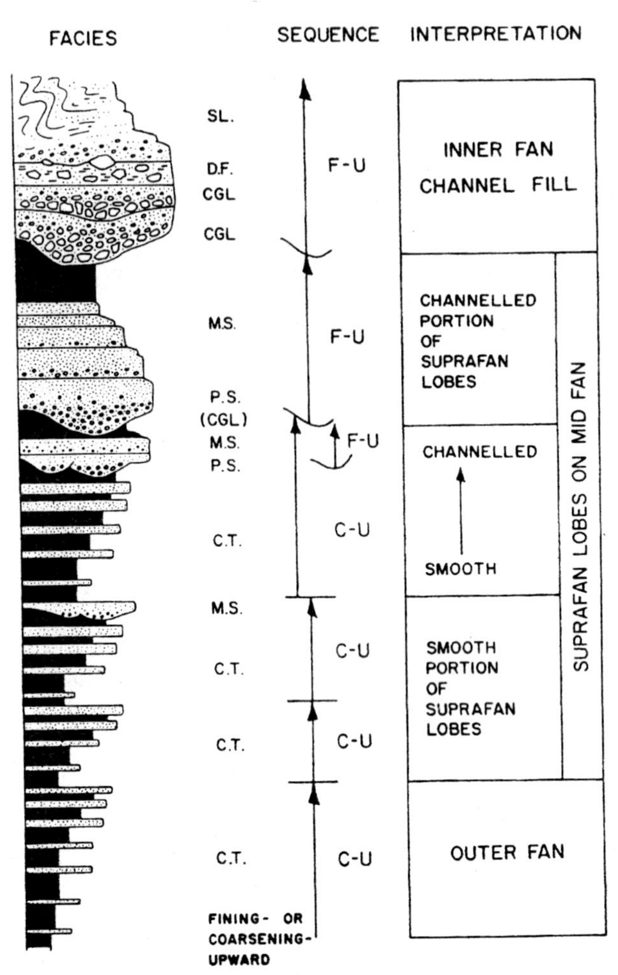

Figure 12.24
Idealized vertical sequence of turbidites produced by submarine fan progradation. From Walker (1979), reprinted by permission of the Geological Association of Canada.

However, the Charlie C sand unit appears to have been deposited also in the proximal to middle part of a submarine fan, but in an altogether different lobe. It is not connected in any way to the Main sand. It is a depletion drive reservoir, and during early production, its reservoir pressure dropped from 600 psi to near 100 psi within two years. Geologic data from 50 development wells on a 700-meter spacing recognized rapid facies changes all across the field, where some sand bodies had cross sec-

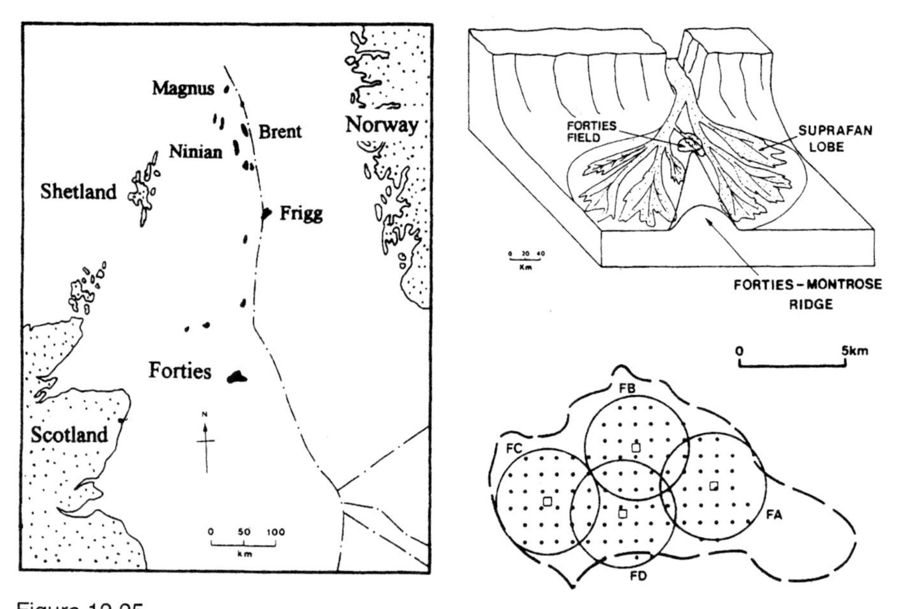

Figure 12.25
Block diagram showing the position of the Forties Field, North Sea, on a submarine fan. Also note platform locations and development well distribution. From Hill & Wood (1980), reprinted by permission of AAPG.

tional areas less than primary well spacing. Many correlations across the field could not be confirmed without the use of pressure decline data and log pattern analyses.

12.12.2 Wilmington Field

In contrast to the Forties Field, the Wilmington Field in California (Figs. 12–27 through 12–29) is situated in the distal part of a fan. The reservoir rocks in the Wilmington Field consist of alternating, thin, turbidite sandstones and shales that are thought to be continuous over great distances. The field is composed of thousands of individual sandstone reservoirs, each with its own water level, its own porosity, its own permeability and its own water saturation (Fig. 12–28, bottom). Volumetric reserve estimates on such fields are difficult because of this tremendous diversity and because of the lack of common water levels in the multitudes of sandstones.

Wilmington is particularly interesting because it underlies the port of Long Beach, California. In the 1950s, withdrawal of oil was causing the land surface to subside at rates of up to 28 inches per year. This is

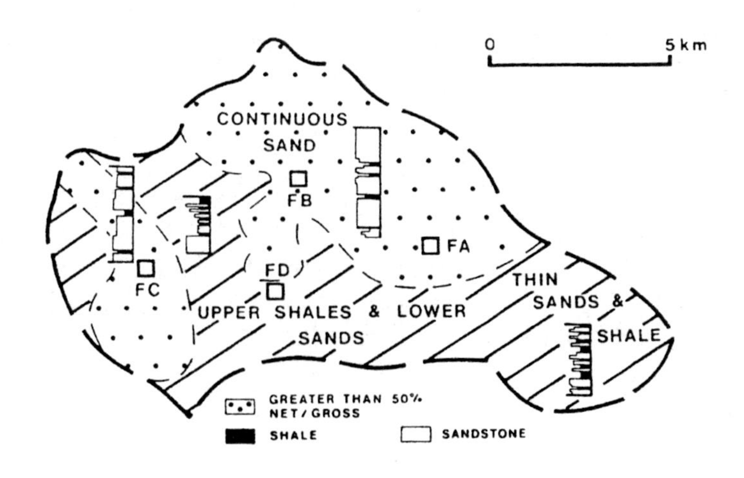

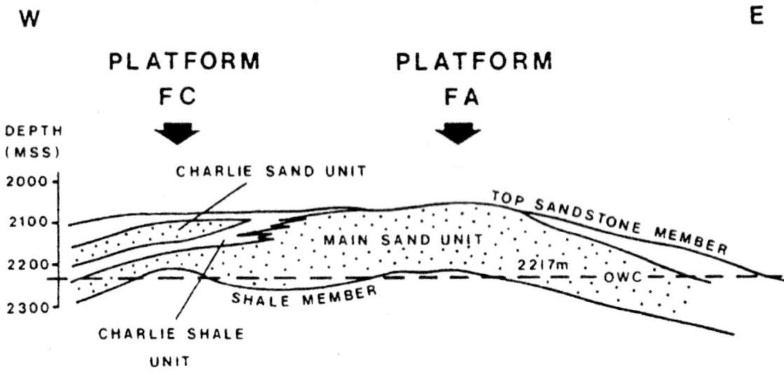

Figure 12.26
Map and cross section views of the Forties Field Main sand unit and Charlie sand unit and their log signatures. Note single water level in Main unit and isolation of the Charlie C unit. From Hill & Wood (1980), reprinted by permission of AAPG.

serious for a port city. Production was also declining seriously. These two problems were solved simultaneously by the initiation of a large-scale water flood project. With the introduction of the project, subsidence of the port ceased and oil production increased dramatically.

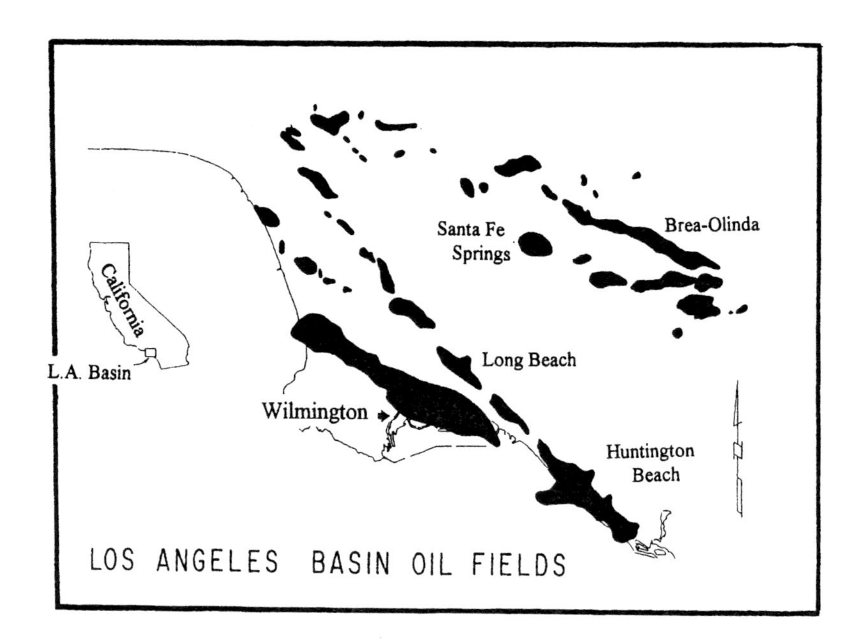

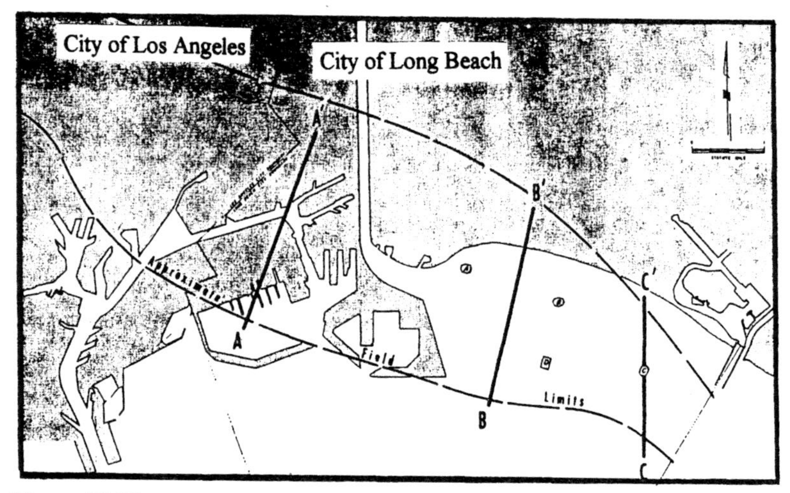

Figure 12.27
Map showing location of the Wilmington Field, California. From Mayuga (1970), reprinted by permission of AAPG.

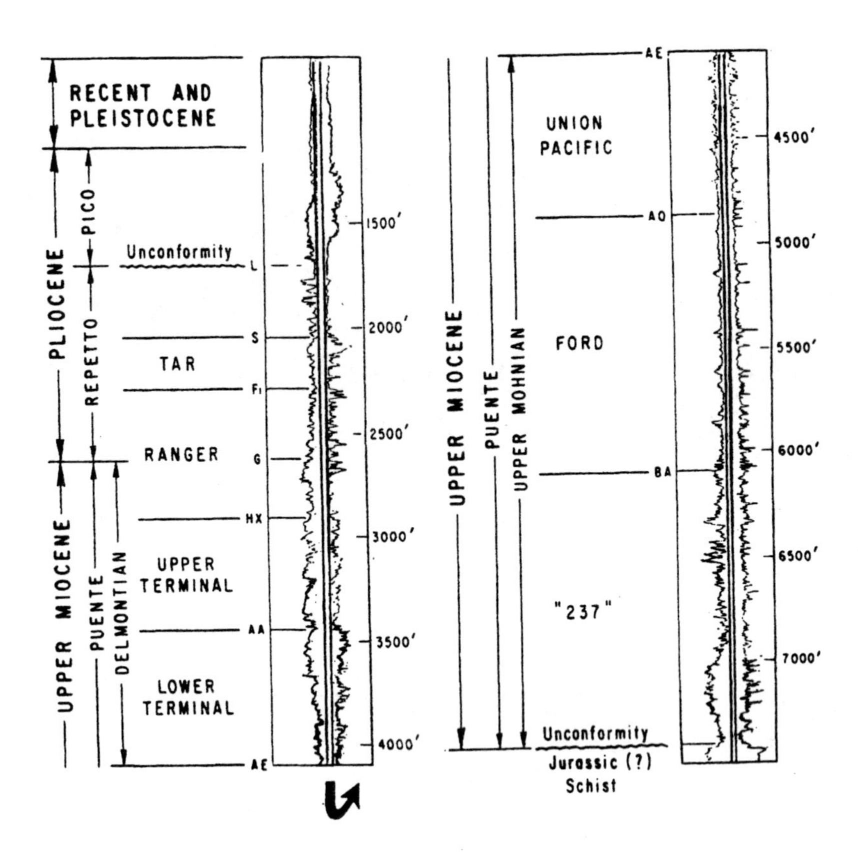

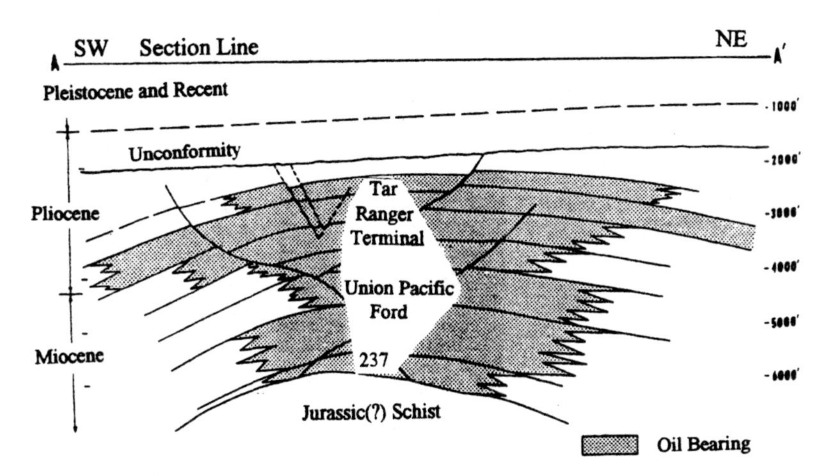

Figure 12.28
Electric log signature and cross section through the Wilmington Field, California. Note the uneven distribution of sandstones and lack of distinct water levels in the cross section. From Mayuga (1970), reprinted by permission of AAPG.

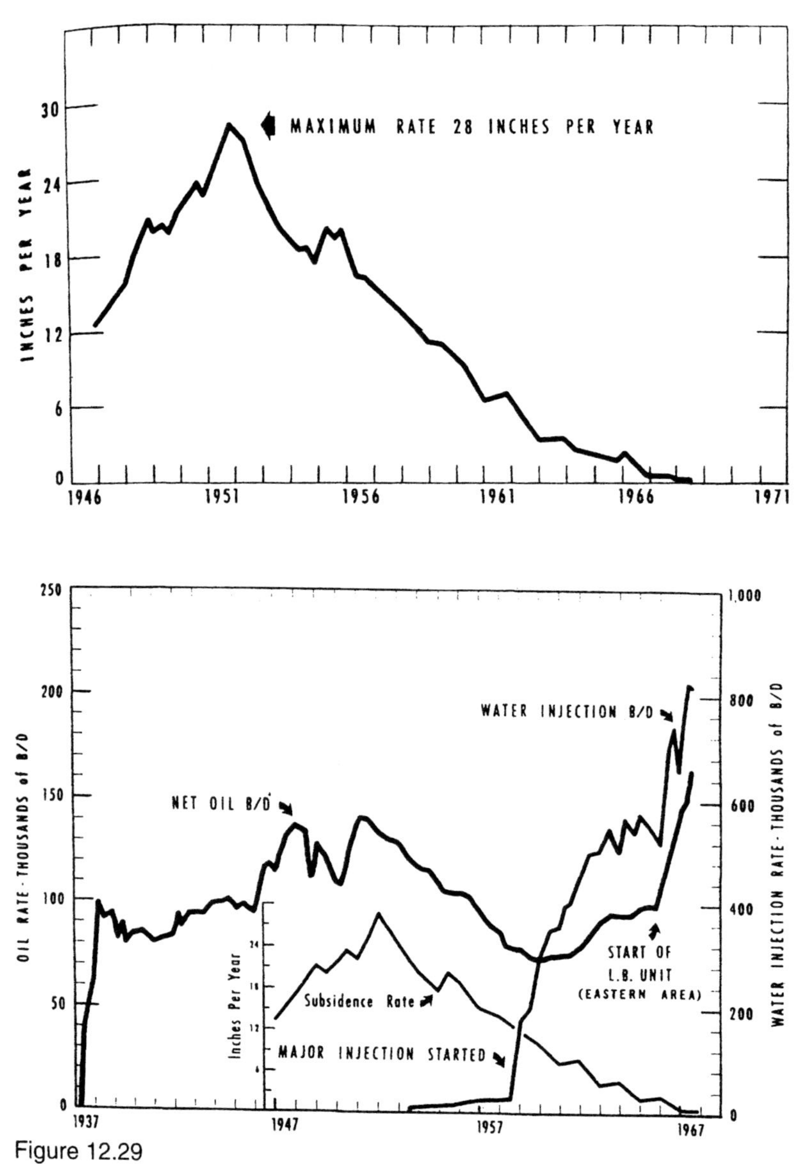

Figure 12.29

Oil production rates, subsidence rates and water injection rates for the Wilmington Field, California. From Mayuga (1970), reprinted by permission of AAPG.

CHAPTER 13

Carbonate Depositional Systems

Four clastic models are important to the petroleum geologist. These models, shown in Figure 13–1, are broadly summarized as:

1. Carbonate ramps
2. Reefs and banks
3. Carbonate slopes
4. Carbonate submarine fans

Of these, carbonate ramps, reefs, and banks are by far the most important, and the majority of this chapter will be spent on them. Good reservoirs in carbonate slopes are rare, but they can make excellent reservoirs. Carbonate submarine fans are by far the least important, as they are relatively rare and reservoir quality is usually poor.

13.1 Carbonate Ramps

Carbonate ramps and the shallowing-upward model constitute the principal mechanism for deposition of widespread, layered (as opposed to lumpy or reef mound) carbonate units. Examples include:

1. The Permian Basin of West Texas
2. The Cambrian-Ordovician Knox-Arbuckle Group of the central United States

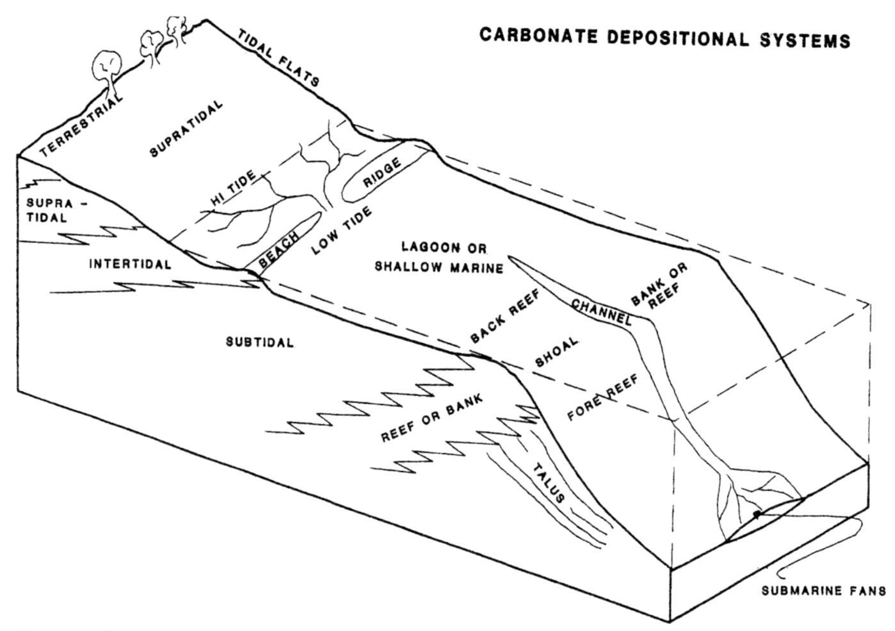

Figure 13.1
Block diagram showing hypothetical composite of several different types of carbonate models.

3. The Mississippian Madison-Redwall Limestone and its equivalents throughout the western U.S. and Canada
4. The Asmari Limestone of Iran, and many other huge limestone units of the Middle East
5. The Mesozoic limestones of Mexico

Although carbonate reefs and banks occur in all of these units, the units are all widespread and layered. In order for oilfields to occur in these units, many have developed some sort of secondary porosity, either diagenetic or fracture porosity.

Carbonate ramp sequences, by definition, are deposited very close to sea level. The following is a brief discussion of porosity development that typically occurs in different environments. Keep in mind that typical porosity development during diagenesis may have little bearing on the quality of the final carbonate reservoir. Carbonates precipitated on the sea floor may have high initial porosities, followed by early diagenesis that closes down virtually all porosity. However, the final reservoir rock may have excellent porosity and/or permeability if the rock is fractured or if it is exposed to meteoric water and karsting. Carbonates often have complex diagenetic histories, including information on primary porosities and early-stage diagenesis tendencies.

Carbonate ramps, represented by supratidal and intertidal areas, commonly have extremely high initial porosities in the form of mud cracks, rip up structures, fenestral textures, and burrows. Early diagenesis can either enhance this porosity or close it down. The prograding supratidal area tends to dolomitize underlying rocks, and the dolomitization process is a process that normally increases porosity to a point, and then porosity decreases. It has been stated that partially dolomitized rocks account for about 50 percent of the world's oil production. Although this number may be too high, this type of rock certainly makes excellent reservoir rocks. The mineral dolomite has a very strong force of crystallization, and a partially dolomitized rock will commonly be composed of a sugary textured group of intergrown rhombohedra with excellent intergranular porosity and permeability. However, if the dolomite is completely recrystallized it may be as tight as cast iron unless it is fractured or karsted.

The subtidal environment is commonly composed of carbonate sand and carbonate mud that initially have porosities similar to terrigenous sediments. However, after deposition, subtidal carbonates may be cemented almost immediately, right on the sea floor. Carbonate muds, many of which were originally composed of aragonite needles, recrystallize to calcite mud and become very impermeable. Thus, subtidal carbonates usually have low matrix porosity, but subsequent fracturing or solution by meteoric water can make them into excellent reservoir rocks.

Because carbonate ramps are deposited very close to sea level, their diagenetic history is often related to sea level changes. A drop in sea level often exposes the entire ramp to subaerial erosion and meteoric water. Exposure to meteoric water can often result in extensive solution of carbonates above the ground water table and reprecipitation below the ground water table. Solution above the ground water table can result in the opening of tremendous fluid passageways in the form of karst features, such as caves and underground open channelways.

A rise in sea level usually causes deposition of a transgressive, carbonate sandstone. Underlying sediments are commonly exposed to sea water, which will tend to precipitate carbonate cements in the pore spaces and destroy porosity.

The point of this discussion is that it is difficult to generalize about carbonate ramps and their porosity-permeability distributions. There are a limited number of porosity types, but their quality varies and every field is different. Electric logs can tell quite a bit about fracture porosity, but they cannot tell much about the diagenetic history of the rock. In order to fully understand the diagenetic history of a field, core samples must be collected and interpreted by an experienced carbonate petrologist.

13.1.1 Asmari Fields

The Asmari Fields in Iran (Fig. 13–2) are dependent primarily on fracture porosity. Fortunately, the very act of folding a rock into an anticline

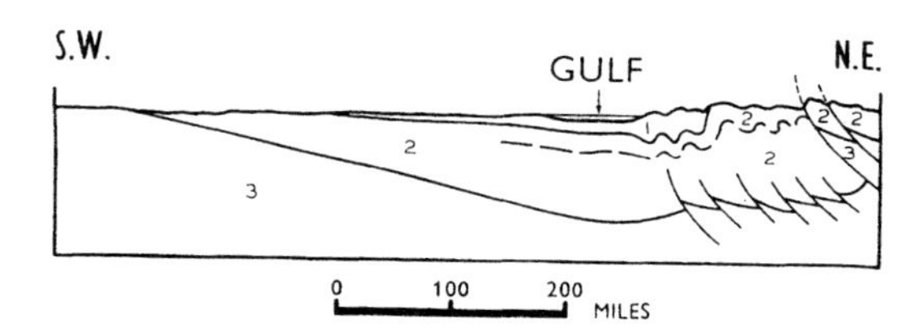

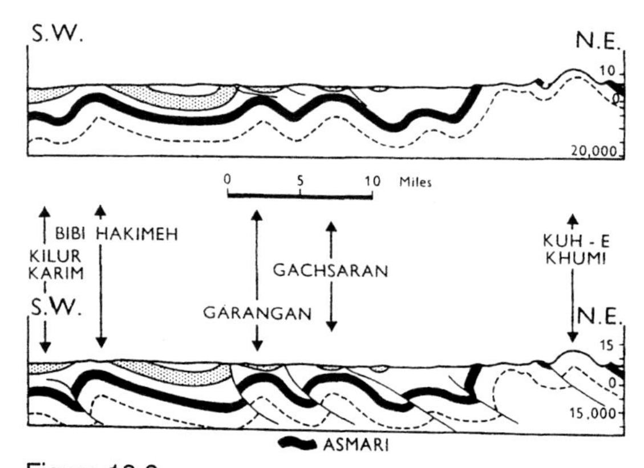

Figure 13.2

Cross sections through the Asmari Fields of Iran. From Hull and Warman (1970), reprinted by permission of AAPG.

commonly forms tension cracks parallel to the principal conjugate shear sets across the crest of the anticline. The core of the anticline is commonly in compression, and the outer part is commonly in tension, and the outer part which forms the crest of the anticline will commonly have well-developed joint sets. This is the case with Asmari and many of the large anticlines of Iran, Saudi Arabia, and throughout the Persian Gulf.

The Gachsaran Field (Fig. 13–2, bottom) has a vertical hydrocarbon column of greater than 6,000 feet and is sealed by a nodular anhydrite layer. Exceptional wells in the field are capable of producing at rates of 80,000 barrels/day, and more than 15 Asmari wells have produced greater than 100 million barrels of oil since discovery in 1928.

13.1.2 Jay Field

The Jay Field of North Florida is an excellent example of a field with a complicated diagenetic history. Smackover clastics and carbonates were deposited in a shallow lagoonal area during the Late Jurassic in an area to the north of Pensacola, Florida (Fig. 13–3). In the field, four principal types of carbonates were deposited, as shown in Figure 13–4. The pelleted micrite was relatively impermeable, the pelleted grainstone had high initial permeability, the fine skeletal micrite had intermediate

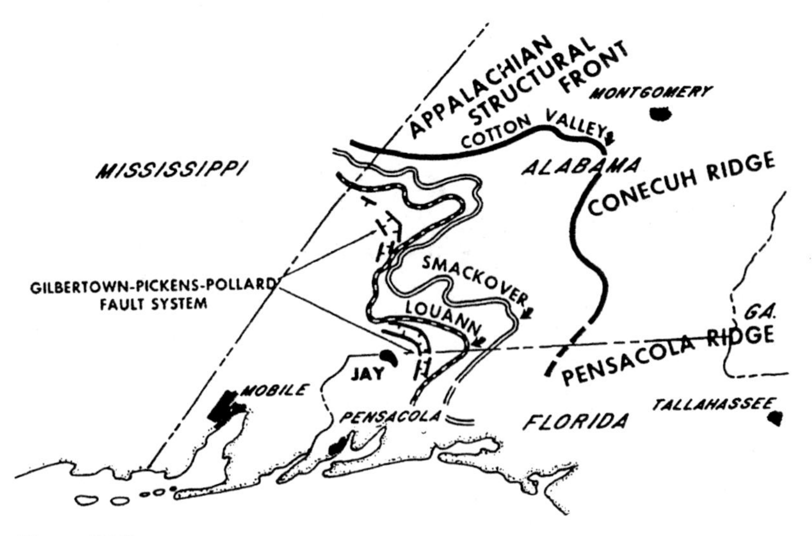

Figure 13.3
Map showing the location of the Jay Field today and its Jurassic geologic setting. From Ottmann et al. (1976), reprinted by permission of AAPG.

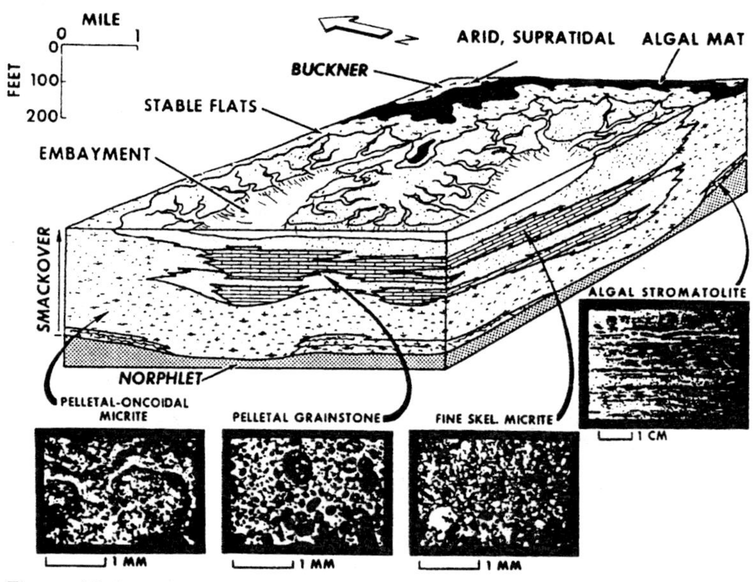

Figure 13.4
Block diagram of the Jay Field showing original carbonate rock types and their settings. From Ottmann et al. (1976), reprinted by permission of AAPG.

porosity, and the algal stromatolite had relatively low porosity. Early diagenesis reduced the porosity in all of these rocks. As the lagoon filled with carbonates and the supratidal environment containing evaporites prograded over the four rock types, much of the underlying rock became partially dolomitized. Finally, a slight tilting of the area caused fresh water to migrate through some of the zones of permeability to develop a moldic porosity, as shown in Figure 13–5. In this case the dolomite formed a cement around the pellets. When fresh water moved through, it dissolved the nondolomitized pellets but left the dolomite cement. Moldic porosity is a sort of reverse porosity. When it occurs, the resulting rock is usually a spectacular reservoir rock. Apparently, two forms of secondary porosity are responsible for the occurrence of the Jay Field. One is the partially dolomitized limestone and the second is the development of the moldic porosity by fresh water. Of the two, the moldic porosity forms the best reservoir rock, as shown in Figure 13–6.

There are two ways to go about developing such a field. It is possible to locate and quantify the porosity using electric logs. Additional development wells can be drilled by simply following porosity and permeability trends. But to develop the field properly and to identify additional

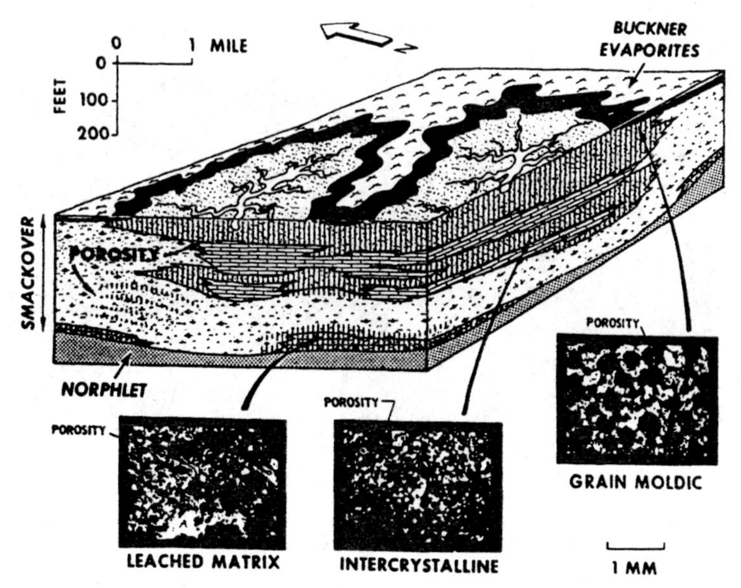

Figure 13.5
Block diagram of the Jay Field showing final porosity distribution (vertical hachures). From Ottmann et al. (1976), reprinted by permission of AAPG.

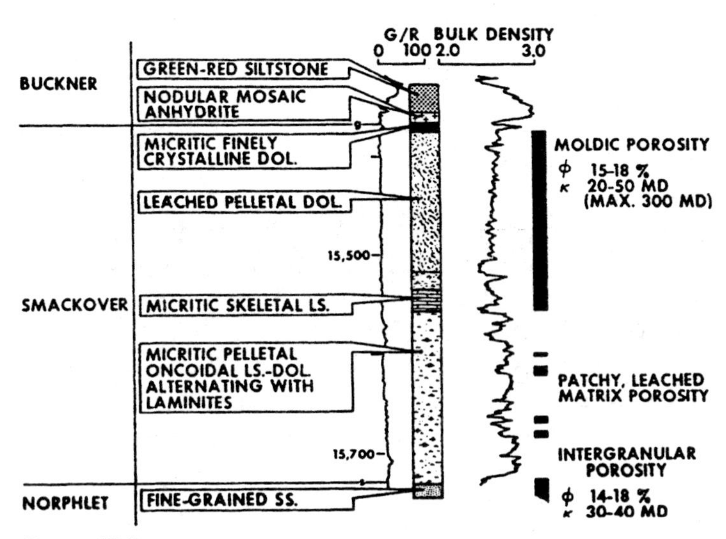

Figure 13.6
Gamma-ray density log showing porosity and permeability distribution as a function of rock type. From Ottmann et al. (1976), reprinted by permission of AAPG.

drilling targets, it is important that cores be taken so that the history of the porosity development can be determined. In carbonates, this cannot be done without cores, and the development geologist should recommend that cores be taken in the development of complex carbonate fields.

These two fields have been singled out as examples not because they are unique, but because they are well-documented in the literature. They are typical examples of layered carbonates where the porosity is developed either by diagenesis or fracturing. The Mississippian limestones of the western U. S. typically must contain reefs or be fractured before they can be considered as reservoir rocks. The same is true for the vast Jurassic and Cretaceous limestones of Mexico.

13.2 Reefs and Banks

According to marine biologists reefs are organic buildups that:

- are attached to the bottom
- have a rigid framework
- are in living position
- thrive in the surf zone

To the sailor, a reef will tear the bottom out of his boat if he runs aground on it. A bank is not rigid, will not thrive in the surf zone, and will not tear the bottom out of a boat.

In the geologic record (for example, on a seismic line) there are many carbonate mound-shaped structures or buildups where it is impossible to tell whether the organic material is attached to the bottom, whether it is in living position, or whether it thrived in the surf zone. The correct term for such a buildup is *bioherm,* but most people call them reefs whether or not they are true reefs.

To the petroleum geologist it doesn't normally make much difference whether the bioherm is a reef or a bank. Both commonly contain oil. Both are organic in origin so they form their own source rock, both are typically mound-shaped and thus form their own trap, and both commonly have outstanding primary and sometimes excellent secondary porosity. They almost always make good reservoir rocks. Not uncommonly they have been exposed to meteoric water and are sometimes karsted, and sometimes they are partially dolomitized. And finally, they can commonly be recognized by reflection seismology. Hence, they make excellent exploration targets.

However, the development of reef fields can be tricky. Reefs commonly grow towards a prevailing wind direction such that to the windward side of the reef crest, there is commonly a reef front that drops off abruptly. A development well drilled seaward from the reef crest is likely to come in structurally low and is likely to be dry unless a fore-reef talus is encountered. As wells are drilled farther and farther into the back reef area, porosity and permeability are likely to decrease markedly. If wells are drilled laterally along the reef crest, channels or small canyons may be encountered.

For the development of large reef fields, the development geologist may want to recommend that additional seismic lines or three-dimensional seismic imaging data be acquired. Porosity and permeability distributions are often erratic and three-dimensional seismic data may help considerably in defining these area. However, there is still a lot of guess work, and some dry holes should be expected. Commonly, small patch reefs are simply one-well fields.

Truly excellent examples of documented reef fields include:

- The Devonian reefs of Alberta—Leduc, Swan Hills, and Rainbow reefs of Alberta, Canada (Fig. 13–7)
- Horseshoe Atol of West Texas (Fig. 13–8)
- The Golden Lane of Mexico (Figs. 13–9, 13–10)

The Golden Lane Field is one of the most interesting fields in the world. The Golden Lane was discovered in 1907 near Vera Cruz, Mexico. Legend has it that in developing the Golden Lane fields, drillers simply drilled (with cable tools) down through the Austin Chalk to near the top of the El Abra Limestone. They drilled very slowly into the top of the El Abra, and when they heard a deep rumbling sound, they all just ran from the well as fast as they could. Initial pressures were so high that all of the initial wells blew out, mud, oil, drill pipe, and all. Some are reported to have blown out pieces of stalagmites and stalactites along with the oil. Collection facilities consisted of throwing up earthen dikes in the creeks, and the oil was simply allowed to flow downhill to the retention dikes (never mind the EPA). Trucks backed up to the lakes of oil, pumped the oil into the trucks, and carried the oil to the refinery at Vera Cruz. Eventually wellhead equipment was installed and the wells were metered. When first metered, the Cerro Azul #4 well was recorded at a rate of 260,000 barrels per day. Many individual wells in the Golden Lane Trend have produced over 100 million barrels of oil.

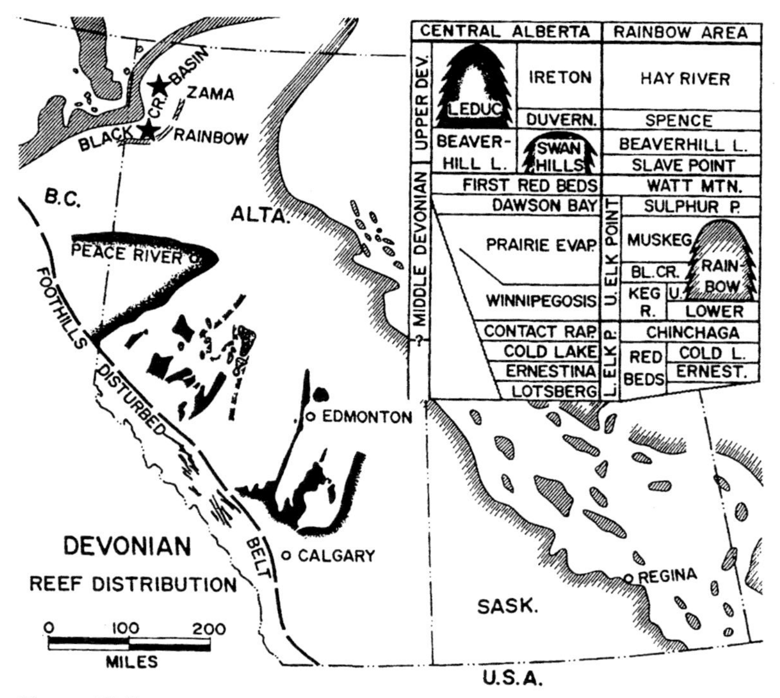

Figure 13.7
Devonian reef distribution of Western Canada. From Barss et al. (1970), reprinted by permission of AAPG.

Figure 13–10 is a schematic cross section through the west side of the Golden Lane atoll. The principal producing horizon is the El Abra facies, a lagoonal miliolid mudstone. It is a micrite with little intergranular porosity. The rare Taninul reef facies has been identified in only two wells. This is also true for the rest of the world as framework reefs comprise a very small percentage of carbonates around the world. Until into the 1930s almost no wells were drilled more than ten feet into the El Abra facies. Until the mid-1930s, the internal fabric or configuration of the reef was virtually unknown.

How did such a fantastic field develop in a miliolid mudstone? The key, once again, is the development of a secondary porosity. Shortly after the atoll formed, sea level dropped, and the whole system was exposed to meteoric water, and the El Abra facies became karsted. Actual caves developed in an otherwise poor reservoir rock. It is believed that many carbonate buildups around the world owe at least part of their excellent reservoir characteristics to meteoric water and karsting.

13.3 Carbonate Slopes

As storms attack the reef crest, large blocks of reef material are commonly carried down the reef front and are piled up at the base of the slope. Not uncommonly a reef breccia builds up at the toe of the slope.

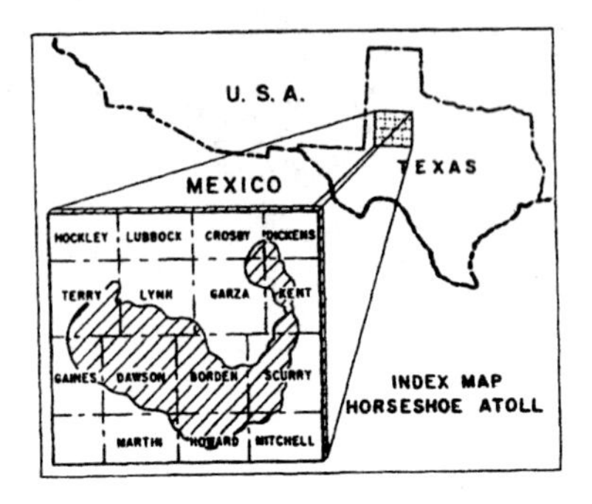

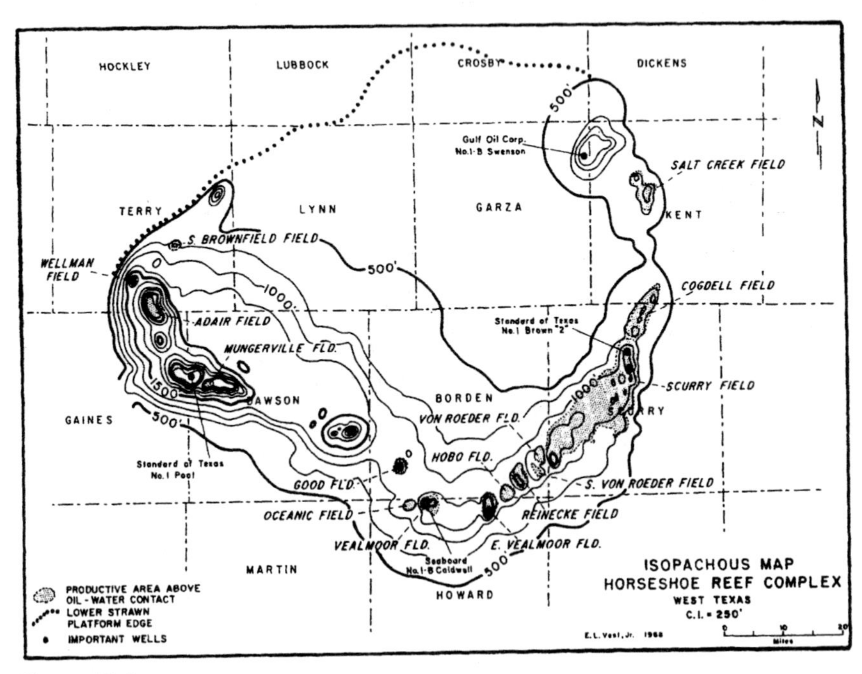

Figure 13.8

Isopachous map of the Horseshoe atoll complex, West Texas. From Vest (1970), reprinted by permission of AAPG.

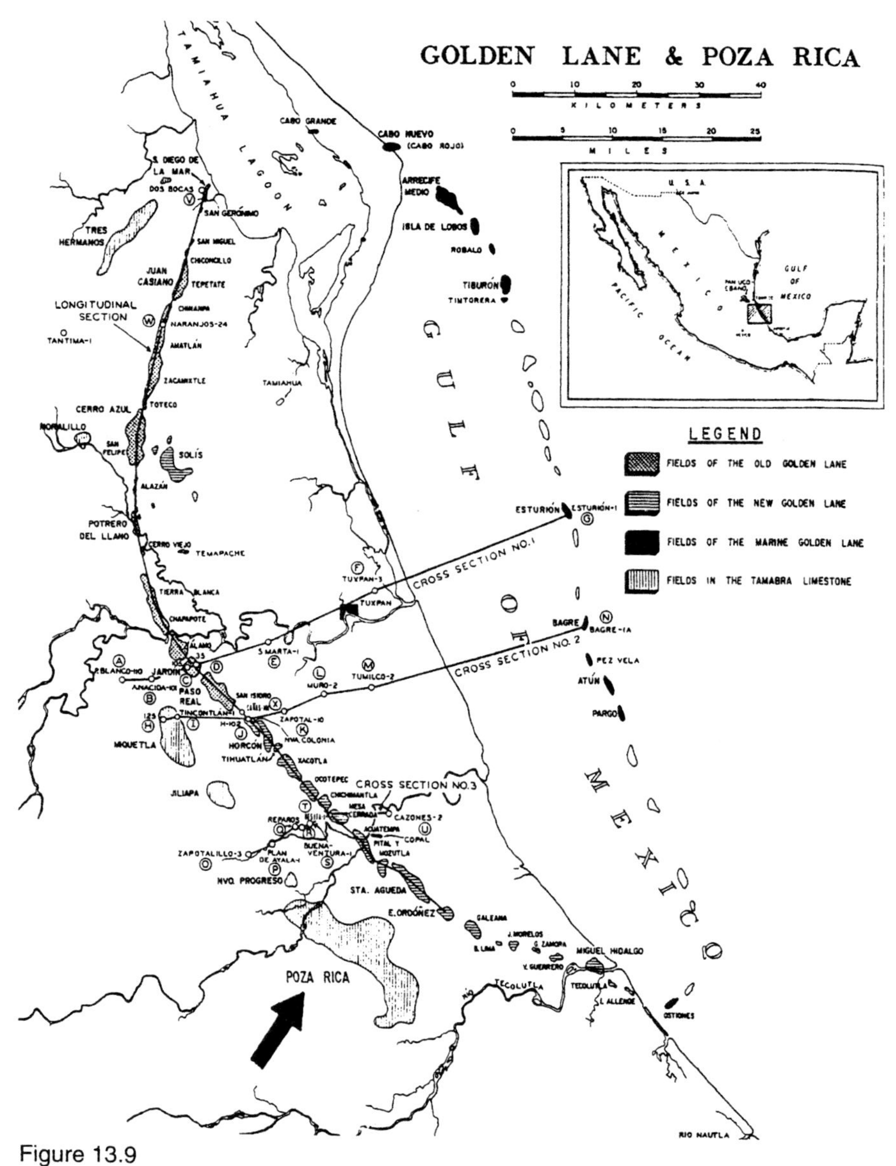

Figure 13.9
Index map showing location of the Golden Lane and Poza Rica, Vera Cruz, Mexico. From Viniegra and Castillo-Tejero (1970), reprinted by permission of AAPG.

The Tamabra facies (Fig. 13–10) is such a facies at the toe of the Golden Lane atoll. In 1938, 30 years after the discovery of the Golden Lane, the Poza Rica Field was discovered in the Tamabra facies to the south of the Golden Lane (Fig. 13–9).

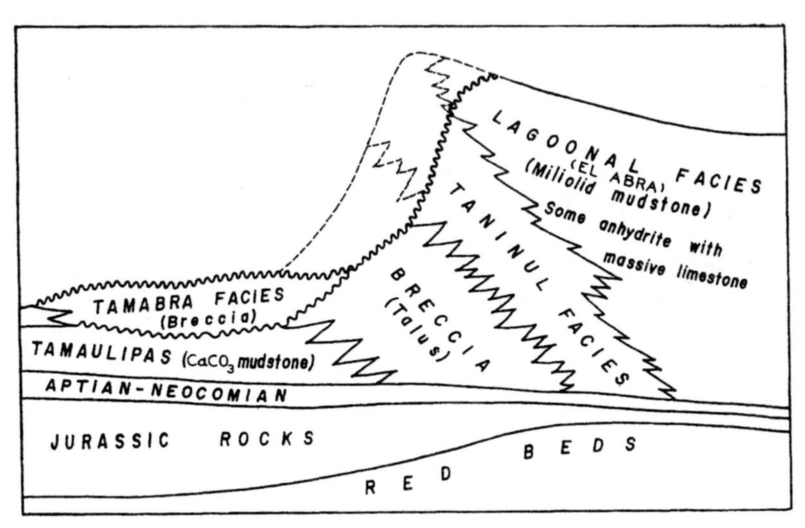

Figure 13.10
Schematic cross section through the Golden Lane atoll-Poza Rica Field, Mexico. From Viniegra and Castillo-Tejero (1970), reprinted by permission of AAPG.

13.4 Carbonate Submarine Fans

In terms of morphology, there is little difference between a carbonate slope and a carbonate submarine fan. The principal difference is that, in a carbonate submarine fan, the turbidites are composed of carbonate clastic material. The sandstone equivalents are calcarenites and the mudstone equivalents are micrites. The calcarenites are commonly very fine grained and well-cemented, and do not generally make good reservoir rocks. The Permian Spraberry-Dean complex of West Texas is an example of a carbonate submarine fan and is described in the next section.

13.5 Example of Combined Carbonate Sequence

The Permian Basin of West Texas is one of the most prolific oil-producing areas of North America. Principal oil production is from ramps, reefs, and banks, with only minor production from carbonate slopes and submarine fans.

The precursor to the Permian Basin, known as the Tobosa Basin, developed as a broad, shallow basin during early Paleozoic time. By middle Pennsylvanian (Strawn) time, the Permian Basin included all of the Delaware Basin, Central Basin Platform, Midland Basin, and the shelf area to the east and north of the Midland Basin shown in Figure 13–11. It was a starved basin, meaning that it had little clastic influx. During middle Pennsylvanian (Strawn) time, carbonate banks built in a horseshoe pattern within the shallow basin to form the foundation for the Horseshoe atoll complex (Fig. 13–8). The Horseshoe atoll complex continued to build as a reef complex through early Permian (Wolfcamp) time, and then became a major oil-producing horizon (Fig. 13–8). At one time (before Kuwait), Scurry County, Texas was noted for having the most millionaires in the world. Interestingly, the reef complex was one of the first reef complexes discovered by reflection seismology, and the first well drilled into the reef, the Gulf No. 1-B Swenson (far northeast corner, Fig. 13–8), was drilled into the only nonproducing reef in the trend. A unique water flood procedure is also used in the Scurry Reef. As a result of initial draining of the central part of the reef, water is now injected into the middle of the reef, and oil is pushed to the edges of the reef, where the reef pinches out into low permeability rock.

At the end of Pennsylvanian time, the Central Basin Platform (Fig. 13–11) was uplifted and became a major ramp complex for all of Permian time. During early Permian (Leonard) time, the Horseshoe atoll complex became covered by carbonate ramps with minor clastics prograding from the northeast, the Abo Reef trend developed on the northwest edge of the Midland Basin, and the Spraberry-Dean submarine fan complex developed in the bottom of the Midland basin. The Spraberry-Dean complex has been referred to as the world's largest uneconomic oilfield. The field occupies a major part of four counties in West Texas, and wells are drilled on a 640-acre spacing. Large fracture jobs are routinely performed as part of the completion procedure on the wells, and returning production is unpredictable and generally marginal at best. The best wells are usually tied to fracture permeability.

Figure 13–11 shows the Permian Basin as it existed during middle Permian (Guadalupe-San Andres) time. By this time, both the Horseshoe atoll and the Abo Reef trends were buried in the subsurface. As sea level fluctuated, complex carbonate ramps formed on top of and on the flanks of the Central Basin Platform. During high sea level stands carbonates were deposited across the top of the platform. During intermediate sea level stands, evaporites were deposited and dolomitization occurred across the top of the platform, and complex oolite banks and reefs formed in shoal areas on the margins of the platform. During low sea level stands the upper parts of the platform were exposed to meteoric water

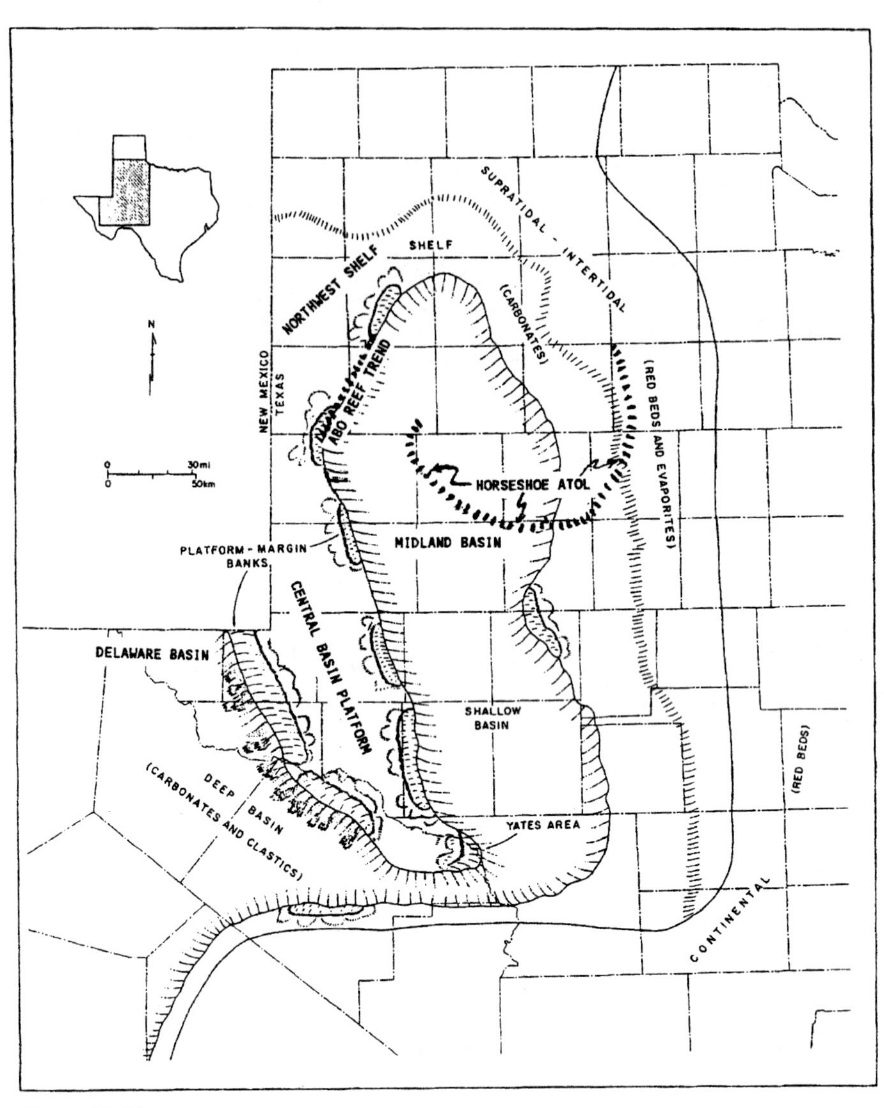

Figure 13.11
Regional paleogeographic setting and facies of Late Pennsylvanian-Early Permian formations of the Permian Basin, West Texas. Modified from Galloway et al. (1983), reprinted by permission.

and erosion. Exposure to meteoric water tended to open up porosity on the highest parts of the platform, while erosion created unconformities. Numerous shallowing-upward carbonate cycles are recorded on the platform, indicating that numerous transgressions and regressions across the platform occurred. In many cases, low sea level stands would open up porosity in rocks across the top of the platform, only to have that porosity closed down by precipitation of evaporites in the pore spaces and as discrete beds during intermediate sea level stands.

As a generalization, the largest and best reservoirs formed on the flanks of the Central Basin Platform, although many fields also occur on the top of the platform. The largest fields occur on the flanks for several reasons:

1. The best original porosities commonly developed in oolite shoal areas and reefs on the flanks of the platform.
2. Draping of subsequent rocks over preexisting reefs commonly formed high areas (anticlines), and the draping sometimes opened up a fracture porosity.
3. These high areas were commonly exposed to meteoric water and the development of secondary porosity.
4. Unconformity traps were commonly developed on the flanks of the platform.
5. Less plugging of porosity by evaporites occurred on the flanks of the platform.

Several points about the Permian Basin are important to the development geologist.

1. As already stated, the diagenetic history of the reservoirs is critical to understanding how the fields developed and to defining the limits of the field.
2. Logs do not tell much about diagenetic history. Cores must be taken and studied in detail.
3. Many of the reservoirs are discontinuous laterally, and serious EOR mistakes can be made if geologic details are ignored.

Detailed correlation of carbonate cycles can be very difficult. Today, major efforts are being made using sequence stratigraphy and flow units to make certain that correlations are correct and that appropriate flow units are defined.

In the past, there have been several water flood projects where water was injected in one well, recovered in another well, and injected back down the first well in a continuous, unending process. It was assumed that the water was helping flush more oil out of the lower permeability beds, but in fact, the water was simply cycling round and round through the most permeable zone which had already been drained. Today, detailed geology is important for water floods, CO_2 floods, and all EOR projects to make certain that the fluids are being injected into horizons where they will do the most good.

CHAPTER 14

Geophysics and Reflection Seismology

14.1 What Is Geophysics?

In the context of most universities, geophysics consists of the following general subdivisions:

1. Gravity—study of the earth's gravitational field
2. Magnetism—study of the earth's magnetic field
3. Electrical methods
   (a) Electrical resistivity
   (b) Magneto-tellurics
4. Seismology
   (a) Reflection
   (b) Refraction
5. Well-log analysis (petrophysics)

Although each of these is an important part of geophysics, in the context of an oil company, geophysics means reflection seismology and/or well logs. The term "geophysical log" is used irregularly, and the science of well logs and their analysis is now more commonly referred to as *petrophysics.*

Gravity, magnetic, and refraction surveys are conducted occasionally by oil companies today, but their use is fairly specific and limited in scope. Reflection seismology is the number one conventional exploration tool used by oil companies today. Throughout the early 1980s approximately 50 percent of the mainframe computer capacity in the U.S. was used for processing reflection seismic lines. Amplitude analysis and three-dimensional seismic imaging are now everyday tools for the development geologist.

14.2 Reflection Seismology

Reflection seismology is the science and business of:

1. Activating a sound (sonic wave) source.
2. Collecting the reflected data that returns from the interior of the earth through arrays of microphones, geophones, or hydrophones, commonly referred to as "jugs" (Fig. 14–1).
3. Processing that data digitally using computers. It includes stacking, several forms of filtering, and migration (discussed later this chapter).
4. Geological interpretation of the data.

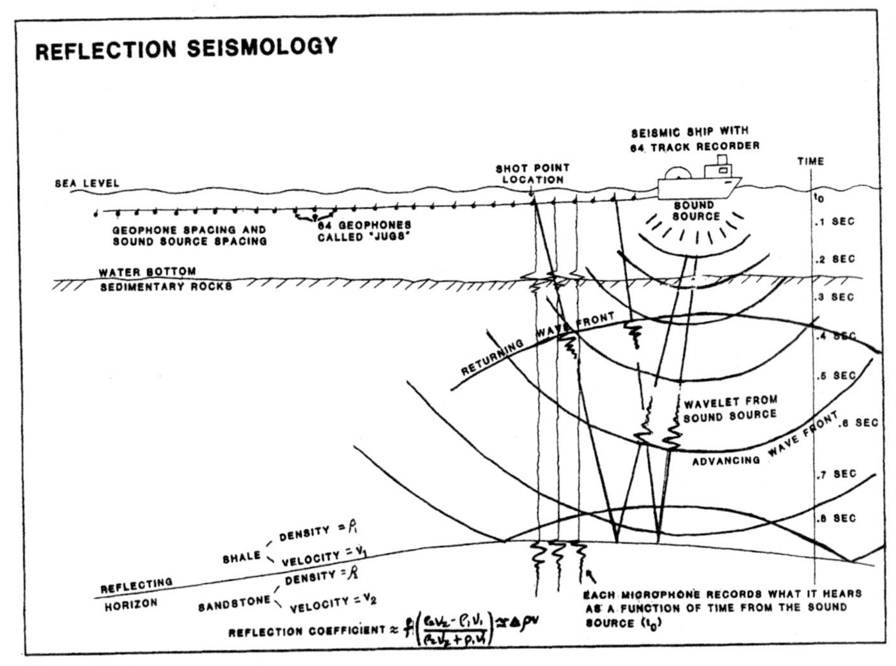

Figure 14.1
Cross section through a typical reflection seismic setup for an offshore situation.

Reflection seismology consists of three principal jobs. The *field crew* collects the data (numbers 1 and 2, above) in the field. This is not a simple process because large geophone arrays are used, and the data is recorded digitally by onboard computers using multichannel recording devices. It is particularly difficult in mountainous terrain where elevation corrections must be made.

The second is the *seismic processor.* This person is a computer programmer or mathematician who looks for ways to enhance the seismic lines after the data has been collected. The seismic processor writes computer programs that manipulate the data in a number of different manners, the most important of which are filtering, stacking, and migration.

The third job category is the *seismic interpreter,* who makes geologic interpretations from the seismic lines. This is usually a combination geologist/geophysicist, someone who knows enough geology to make geologic sense of the data, and who knows enough geophysics to know which geophysical data is believable and which is not.

14.3 First Seismic Lines

In the first seismic lines that were developed, arrays of geophones (microphones) were spaced outwardly from a sound source, usually a dynamite charge. Later, sound sources included truck-mounted devices similar to that shown in Figure 14–2. As the sound source was activated at t_0, the geophones simply recorded, either optically or with a pen, any energy that returned to the geophones as a function of time. Shear waves, compression waves, and ground roll waves were generated by the sound source. Direct travel path arrivals were filtered out, and the energy recorded at the geophones was (and is) derived principally from reflections off contrasting velocity and density horizons in the earth. The vertical scale on a seismic line is two-way travel time, the time it takes for sound to travel from the source to the reflector and back to the geophone (Fig. 14–2). Because two-way travel time to distant geophones (*stepout*) is greater than that to close geophones, reflections from a horizontal reflector in the subsurface appeared as a hyperbola to the geophone array (Fig. 14–2), if no stepout correction was made. The process simply did not work very well until some correction could be made for stepout and data from multiple shots could be incorporated into a single, composite line. Also, energy from a single shot is rapidly diminished in the subsurface and returning signals were hidden within background noise.

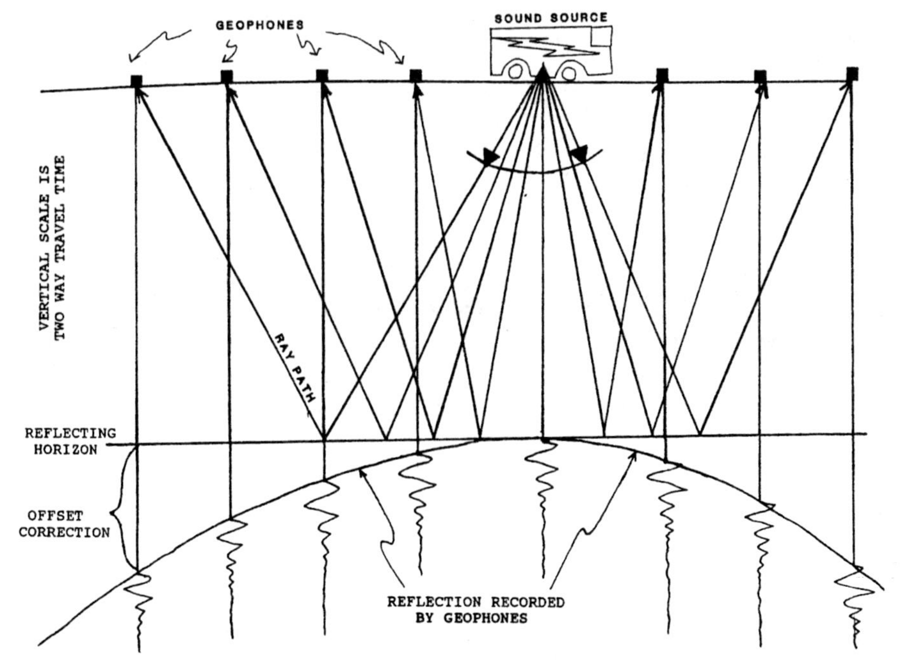

Figure 14.2
Reflections received from a single source and single, flat reflector showing reflection offset as a function of travel time.

Advances in reflection seismology have been closely tied to advances in computers. In order to combine data from several shots, the returning energy is *digitized*—the amplitude of the returning signal is sampled and recorded in the computer as a linear array of numbers. This array of numbers is recorded for each geophone location. The process is important because the linear array of numbers can be manipulated mathematically. Stepout corrections can be made easily by moving the numbers according to the correct velocity hyperbola. But the real value of digitization and computer technology is that data from many shots can be corrected for stepout and added together, or *stacked,* to create a composite line derived from numerous shots.

14.4 Working with the Data

14.4.1 Stacking

Stacking, the most important process in reflection seismology, is the process whereby digitized data from multiple shots and multiple geophones (ranging from 12 to 1064, commonly 24) are added together, or stacked, mathematically to form single traces. Without stacking, weak data are hidden in background noise. By combining data from multiple shots and multiple geophones, random background noise, sums to near zero. Data derived from real reflections using multiple geophones should consistently be either positive or negative and will be clearly enhanced when summed during stacking. Because of stacking, good reflections can be seen from 30,000 to 40,000 feet below the surface in some parts of the world.

14.4.2 Sound source

For very shallow investigations, students use a hammer hitting a steel plate. Dynamite and vibroseis are used commonly today. Vibroseis is generated by a truck that lifts off the ground onto pads that then vibrate at changing frequencies. For shallow offshore investigations, sparker lines are run. The source for a sparker line is a large spark plug that vaporizes the water, causing the water to implode and create a signal. For deep offshore investigations, large metal plates are pulled apart quickly, and the imploding water gives a strong and consistent signal.

14.4.3 Reflection coefficient

As compression waves travel through the earth, part of the waves are reflected back as a function of:

$$\text{reflection coefficient} = \frac{\rho_2 v_2 - \rho_1 v_1}{\rho_2 v_2 + \rho_1 v_1} = \Delta\,\rho v$$

where:

ρ_1 = density of the rock overlying the interface
v_1 = sonic velocity through the rock overlying the interface
ρ_2 = density of the rock underlying the interface
v_2 = sonic velocity through the rock underlying the interface

14.5 Convolution and Deconvolution

14.5.1 Convolution

The reflection process is a convolution process (Fig. 14–1), meaning that the whole wavelet, whatever its signature, is reflected back at the interface. Wavelets change as they travel through the earth. Certain frequencies (especially high frequencies) are naturally filtered out by the earth, and many trailers are picked up as a result of multiple shallow reflections.

If each reflecting interface is thought of as a spike, the reflecting process (convolution) means that each spike reflects back the entire wavelet, but reflects only a small amount of the original energy. In the subsurface there are thousands of lithology changes (spikes), each of which reflects back the full wavelet, but only a small part of the energy. Thus, the energy that reaches the geophone is a mass of jumbled, convolved energy derived from all different reflecting horizons.

A stacked seismic line shows all of this energy (Fig. 14–3). Fortunately, the energy in the wavelet is usually front-end loaded, so the first bursts of energy arriving at the geophone from particular horizons will be derived from a single, strong reflector. A strong reflector will show up well on a seismic line, but it is likely to have many trailers, and information from immediately below the strong reflector will be masked or hidden.

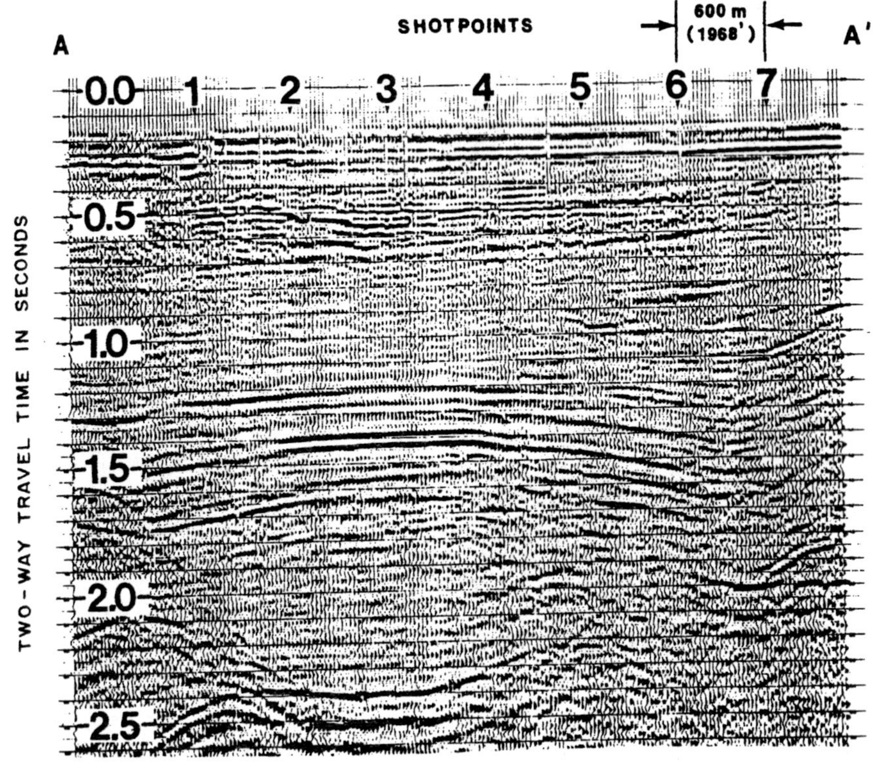

Figure 14.3
Standard seismic section with strong reflection between shotpoints 2 and 4 at 1.4 seconds. Modified from Schramm et al. (1977), reprinted with permission of AAPG.

14.5.2 Deconvolution

The deconvolution process is a filtering process involving Fourier transforms and other mathematical procedures to filter out the wavelet in an attempt to return to a single spike from each reflecting horizon. If the signature of the wavelet is perfectly known, the deconvolution process will convert the jumble of wavelets to a series of spikes that represent reflecting horizons. But, as already mentioned, since the signature of the wavelet itself changes, the deconvolution process is never perfect.

Under certain circumstances deconvolution can be pretty good, however, and it now allows seismologists to study the absolute value of amplitudes from individual, specific horizons. Because many of the trailers have been eliminated (or at least reduced) by this process, the seismologist may also look with confidence at horizons immediately below strong reflectors.

14.6 Bright Spots and Dim Spots

For any specific horizon in the subsurface, density and velocity determinations can be made from nearby well logs. This is important because expected reflection coefficients can be calculated from well logs for any specific horizon in the subsurface. In fact, synthetic seismic traces are commonly constructed from well logs, then overlain on a seismic line to see if they match (Fig. 14–4). Significant deviations from the expected reflection coefficients are often caused by the presence of hydrocarbons in the form of unusually strong reflections (bright spots), as shown in Figures 14–5 and 14–6 or unusually dim reflections (dim spots). Typically, at shallow depths (0 to 7,000 feet) a gas-bearing zone will show up as a bright spot because the addition of gas to a sandstone causes an increased velocity and density contrast across the shale-sandstone interface.

At greater depths (typically 10,000 to 15,000 feet) the addition of gas to a sandstone can cause the density and velocity contrast to be diminished across a sand-shale interface. When there is no difference between the density-velocity products across an interface, the reflection coefficient is zero. When this happens, reflections from water-bearing sandstones may be normal, but through the gas-bearing part of the sandstone there may be no reflection at all, resulting in a dim spot.

Seismic amplitude analysis is very good for the identification of gas-bearing zones in sand-shale sequences. Unfortunately, limestones and very small amounts of free gas (as little as 2 percent in a sandstone) can give very strong reflections. The determination of whether a bright spot

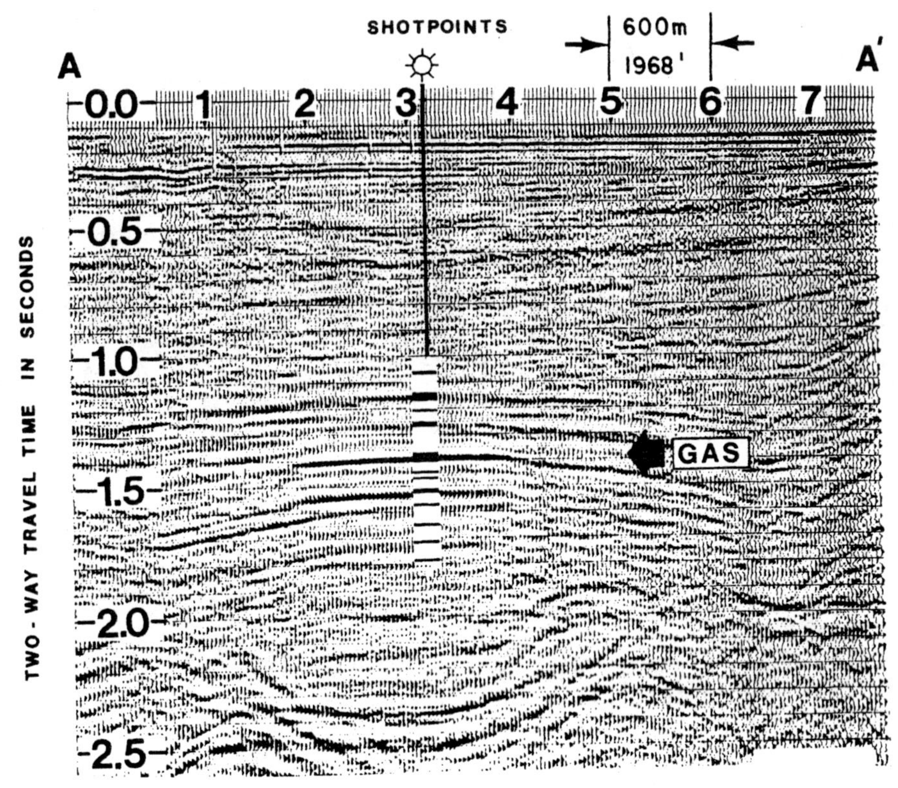

Figure 14.4
Synthetic seismic trace lithology log overlay on deconvolved (wavelet processed) data. Lithology log is derived form sonic and density log data. Modified from Schramm et al. (1977), reprinted with permission of AAPG.

is caused by small amounts of free gas, calcite cement, or a significant gas reservoir is often difficult. One test that has been considered important is to determine whether the limits of the bright spot conform to structure and potential fluid levels. If the bright spot limits conform to a water level on a particular structure (Fig. 14–7), the probability that the bright spot is caused by gas is good. Bright spots or dim spots that do not conform to structure or do not show a common water level on different lines should be regarded with suspicion. On the other hand, bright spots that do not conform to structure could possibly be caused by gas in a stratigraphic trap. Because oil and water are both liquids with a small density differential, identification of oil based on amplitude analysis is difficult, but can be done on occasion.

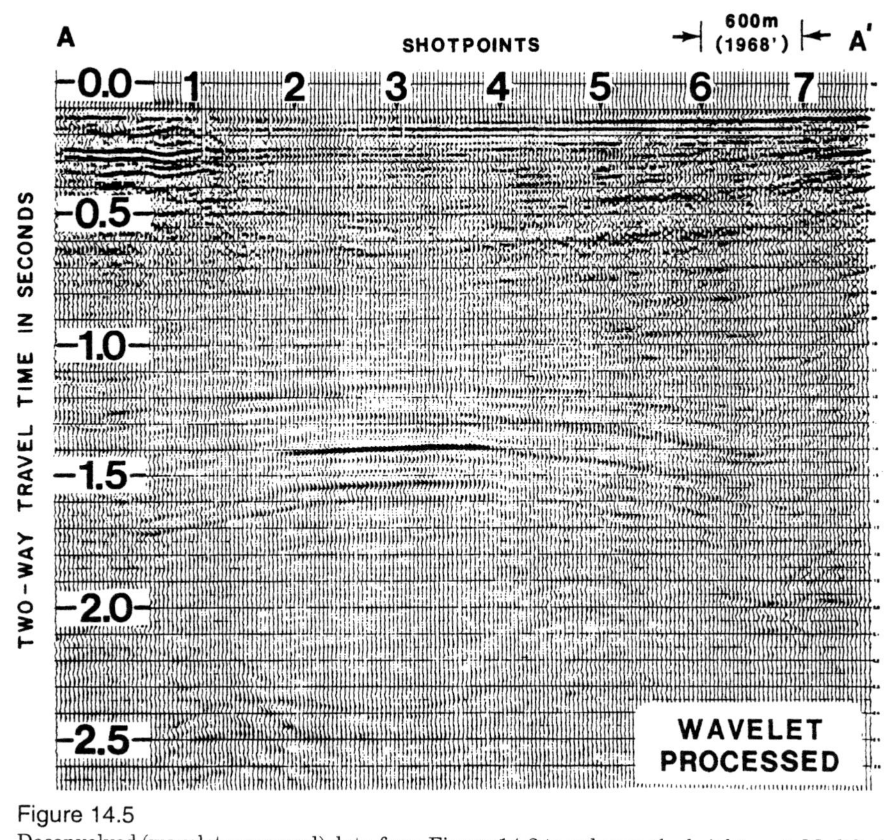

Figure 14.5
Deconvolved (wavelet processed) data from Figure 14-3 to enhance the bright spot. Modified from Schramm et al. (1977), reprinted with permission of AAPG.

14.7 Bed Thickness

Theoretically, deconvolved data can be used to determine bed thickness, because reflections from the top of the bed should deflect (kick) one direction (+), while reflections from the base should kick the other direction (–). Vertical resolution is a function of depth and frequency resolution. With higher frequency, bed thickness resolution is greater.

Unfortunately, the earth tends to filter out high frequencies, and vertical resolution using seismic data is not very good. At 5,000 feet in the subsurface, it is unlikely that vertical resolution is any better than 50 to 70 feet, and it becomes worse with increasing depth.

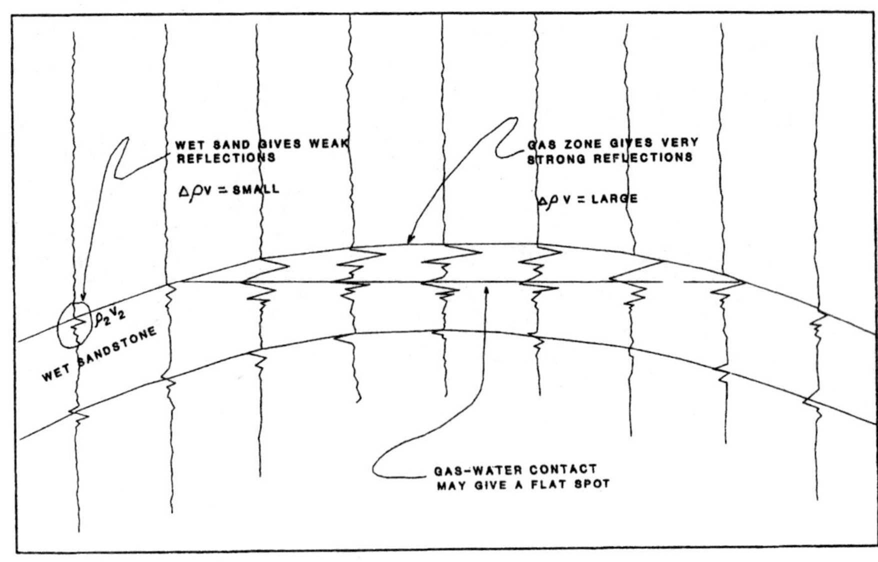

Figure 14.6
Schematic to show amplitude increase (bright spot) through gas-bearing part of the sandstone and flat spot caused by the gas-water contact.

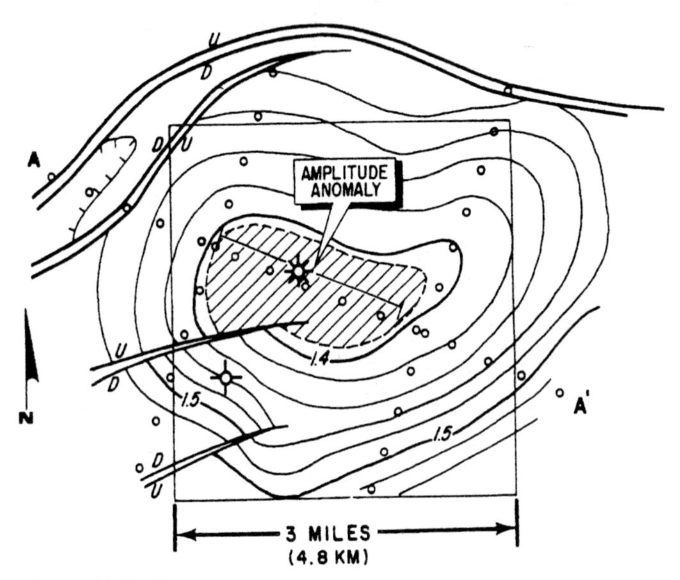

Figure 14.7
Seismic map showing amplitude anomaly (bright spot) that conforms to structure. That is, the apparent gas-water contact is consistent on both sides of the potential reservoir. From Schramm et al. (1977), reprinted with permission of the AAPG.

14.8 Fresnel Zones

To the development geologist, one of the principal uses of seismology is to identify limits, particularly downdip limits, of both tested and untested reservoirs. The edges of bright spots and dim spots are often difficult to define, and even when defined, they may not correspond precisely to the limits of the reservoir because of Fresnel zones. The topic of Fresnel zones is beyond the scope of this text, but we mention them here to emphasize that the edge of a bright spot may not correspond precisely to the edge of a reservoir.

14.9 Migration

14.9.1 Two-dimensional migration

On a conventional seismic line, dipping beds, including bright spots, are plotted in the wrong place (Fig. 14–8) because, for unmigrated data, it is assumed that the reflection occurred directly below the trace. Because sound goes out spherically, reflections can come from anywhere, and for

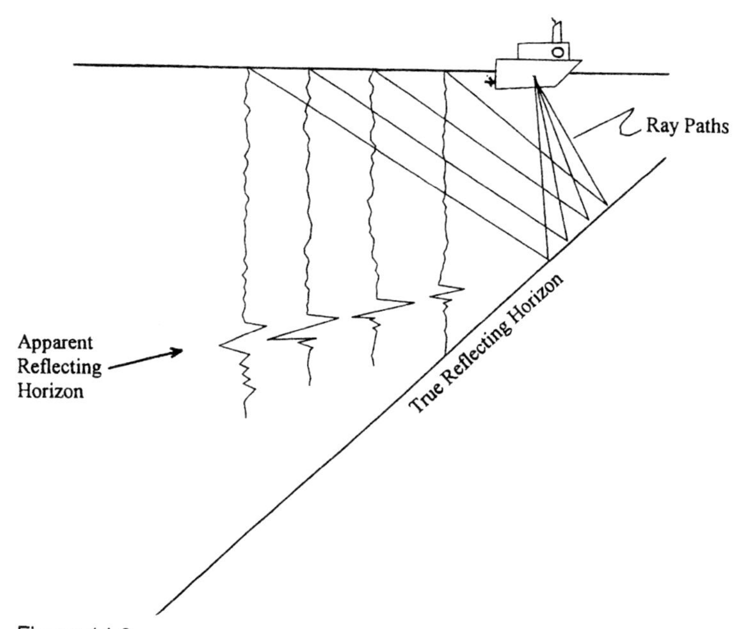

Figure 14.8
Cross section through an exaggerated dipping reflector showing that the derived reflections are plotted at incorrect locations.

dipping beds, the majority of the returning energy comes not from the trace location, but from updip of that location. If the seismic line is oriented perpendicular to the strike of the beds, then the energy comes from updip locations on the seismic line.

The process of migration is done by a computer program that mathematically moves every digitized data point on every trace laterally into adjacent traces along a diffraction curve, representing all locations from which the data could have come. Incorrectly migrated data should be random and should filter out, but correctly migrated data should enhance in much the same way that stacking causes real reflections to emerge from noise. The process works reasonably well as long as velocities are correct and as long as the seismic line is oriented perpendicular to the strike of the beds.

If the seismic line is not perpendicular to the strike of the beds, the returning energy comes from an area not directly beneath the seismic line, but from updip (Fig. 14–9). This is called *sideswipe*. The migration is correct in the context of the two-dimensional seismic line, but it is

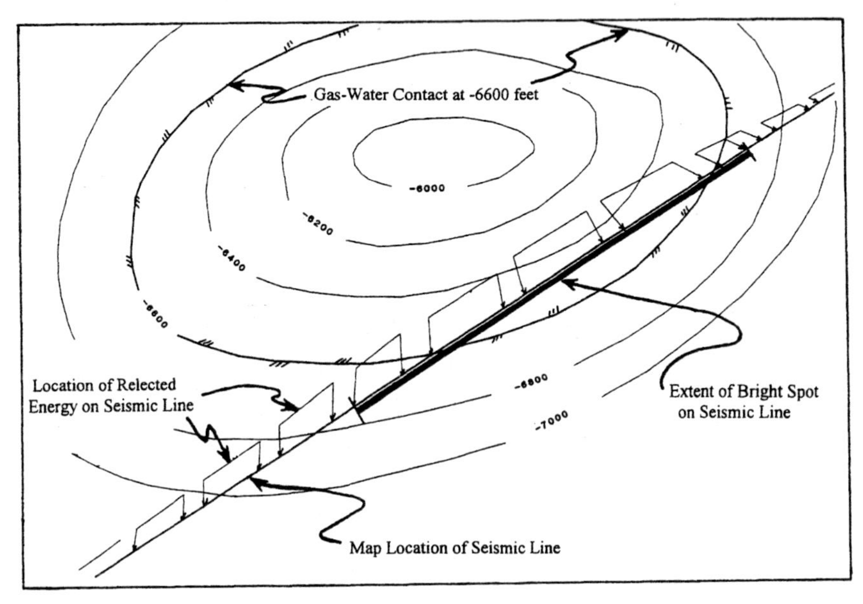

Figure 14.9
Map view of an anticline and a seismic line that sideswipes or does not pass over the crest of the structure. Energy shown on the seismic line is derived from subsurface locations that are located updip of the location of the seismic line.

not three-dimensionally correct because the energy did not come from directly below the seismic line.

14.9.2 Three-dimensional migration

Three-dimensional seismic imaging is one of the development geologist's most important new tools. It is expensive, but the degree of precision that can be achieved is so great that in many cases it more than pays for itself through better placement of development wells and elimination of dry holes.

In three-dimensional migration, data from individual seismic lines are migrated not just two-dimensionally, but three-dimensionally into crossing seismic lines. Three-dimensional diffraction cones are used for the migration algorithm. Data from intersecting lines can be migrated into areas where no seismic lines exist at all. The procedure clearly requires large computer storage capacity and fast manipulation capabilities.

Three-dimensional seismic imaging has several properties that make it extremely valuable to the development geologist. One is the precision to which faults can be determined. Two-dimensional imaging has always had problems with faults, which commonly have strong diffractions in dip lines (perpendicular to fault strike), and disappear altogether in strike lines. For two-dimensional seismic imaging a fault precision of +/–500 feet is considered good at 7,000 feet in the subsurface. Using three dimensions can reduce this error by at least an order of magnitude. For planning of development wells, accuracy is extremely important.

Three-dimensional seismic imaging is also able to detect very small faults that are not normally detected through two-dimensional seismic analyses or through well-log fault cut interpretations. When three-dimensional analyses are run on old fields, new faults are almost always detected.

An important feature of three-dimensional seismic imaging is that time slices (Fig. 14–10) can be converted almost directly to structure contour maps. Time slices can also be taken at different levels to create fault plane maps and to observe fault migration (due to dip) with depth.

Another extraordinary feature of this technique is that data can be projected into areas between lines where no data was originally collected. New synthetic lines can be generated that run between the original lines, and they need not be horizontal or vertical. They can be taken along bedding planes (seiscrop) so that sedimentary facies, such as meandering stream sequences (Fig. 14–11) or barrier island sequences, can occasionally

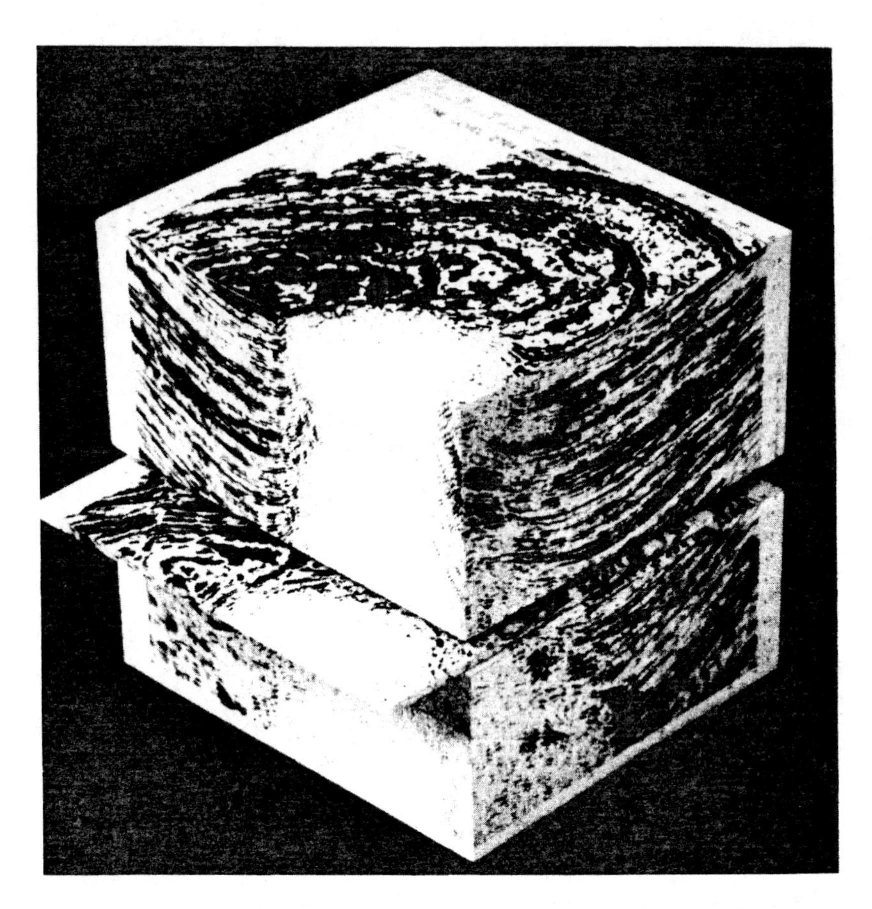

Figure 14.10
Three-dimensional data volume showing a Gulf of Mexico salt dome and associated rim syncline. The diagram also shows how time slices can be pulled from the data sets. From Brown (1988), reprinted with permission of AAPG.

be observed. Also, stratigraphic traps can occasionally be observed that could never have been identified without this extremely powerful new tool.

Finally, one of the most spectacular new applications is in tertiary recovery techniques. In onshore situations, geophones are planted in the ground and conventional three-dimensional seismic analysis is run. Steam is then injected or a fire flood is advanced, and the seismic imaging is run again. The heat caused by the injection of steam or the advance of the fire flood causes a change in the acoustic impedance such that the advance of the steam or fire flood in the subsurface can be monitored precisely (Fig. 14–12). This information can be extremely important for determining the effectiveness of an EOR project where the permeability distribution of the reservoir is irregular. The procedure does not work well in offshore situations, because it is difficult to repeat statics on different runs. That is, it is impossible to place the hydrophones (pulled behind boats) in precisely the same positions during different runs, although rapid advances are being made in this area.

Figure 14.11
Seiscrop section from the Gulf of Thailand showing a meandering stream channel. Modified from Brown (1988), reprinted with permission of AAPG.

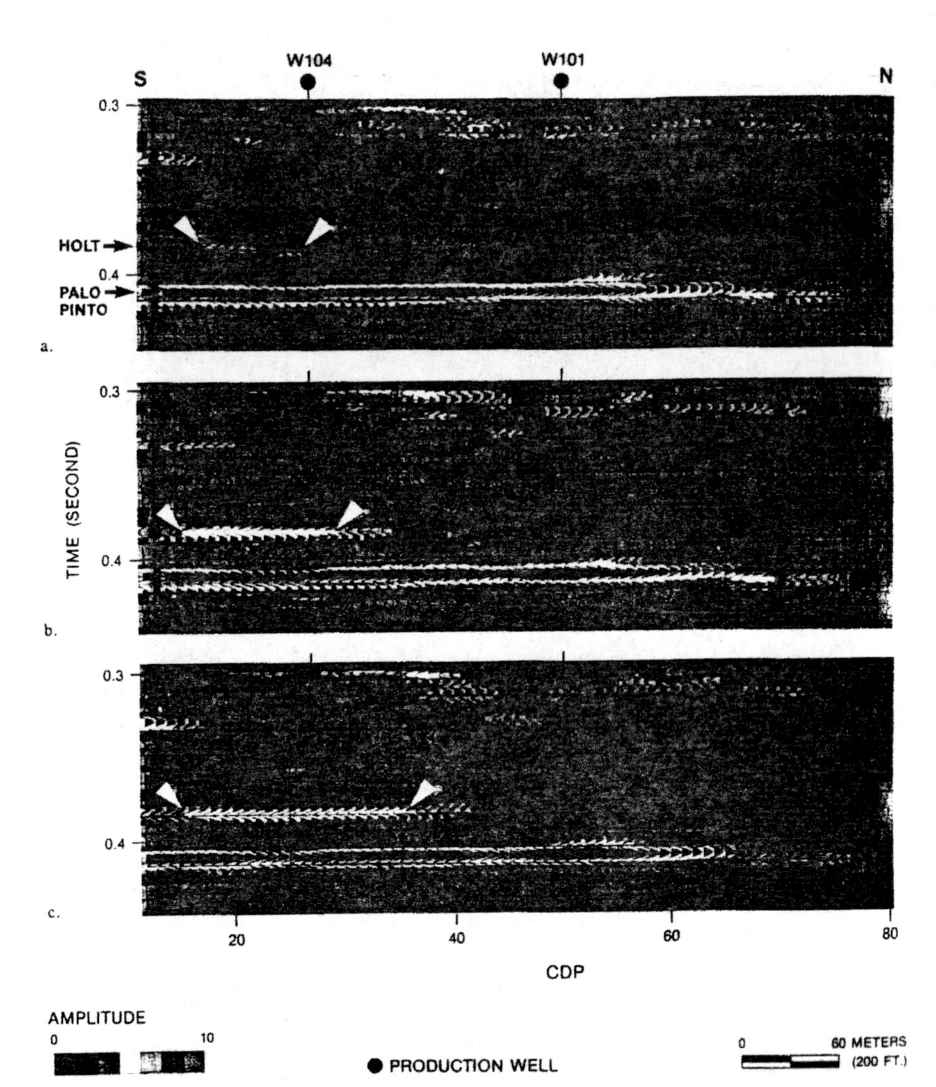

Figure 14.12
Three seismic lines through a fire flood showing preburn, midburn, and postburn conditions. From Brown (1988), reprinted with permission of AAPG.

14.10 Vertical Seismic Profiling (VSP)

Vertical seismic profiling and reverse seismic profiling are new tools that are being used with varying degrees of success in the industry. Most are being used in exploration situations, but there are situations where they might be useful to the development geologist. In particular, reverse VSP can be used as a predictive tool for identifying rocks ahead of the drill bit by using the drill bit as an acoustic source and geophones on the surface as receivers.

Figure 14–13 shows the geometry and resulting returning energy for a typical VSP arrangement. The slope of the resulting curves is a

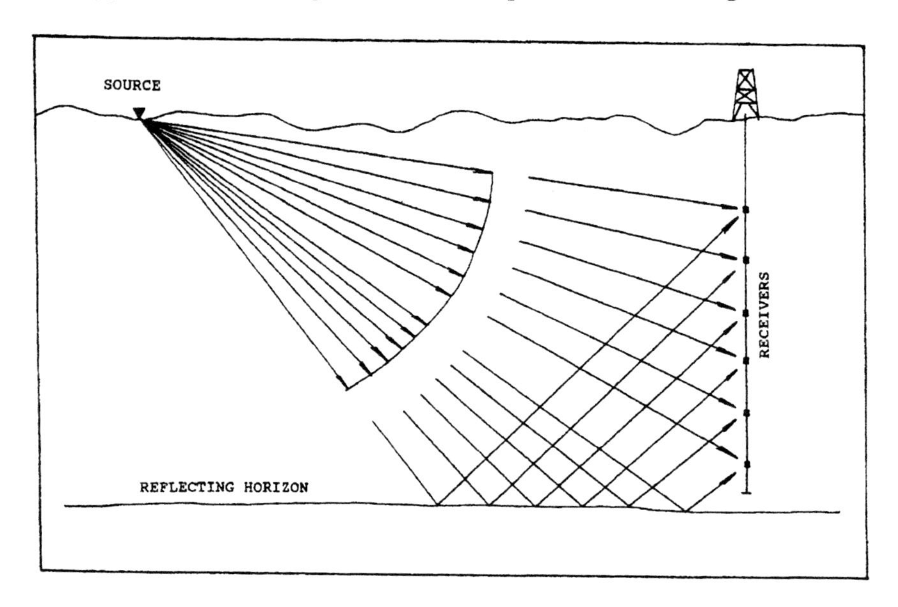

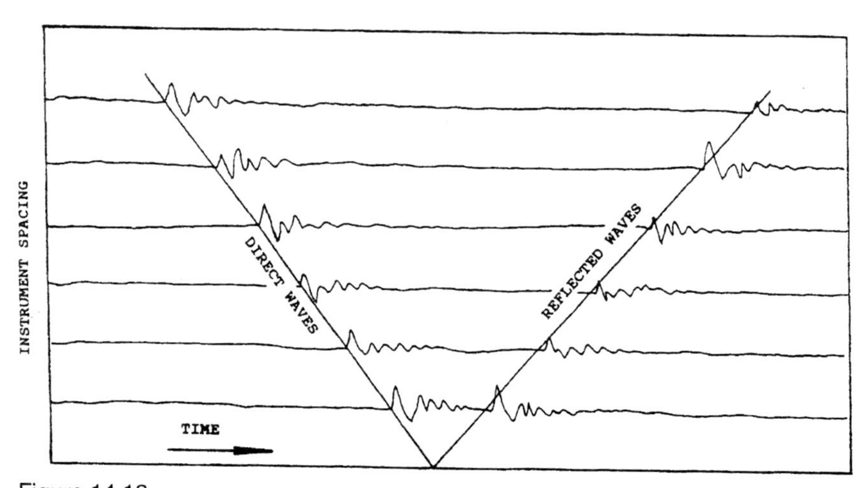

Figure 14.13
Geometry and resulting energy for a typical vertical seismic profile (VSP) arrangement. Slope of the line for the direct waves gives velocity. Asymmetry gives dips and convergence point gives depth to reflecting horizon.

function of the rock velocity, and the symmetry is a function of the dip of the beds. One of the unique features of the technique is that it becomes possible to distinguish between true reflectors and multiples.

Vertical seismic profiling can also be used to look horizontally to identify the location of faults or the edges of a channel. It can be used like an ultra-long spaced sonic log to locate the edges of salt domes or other steeply dipping surfaces that are normally difficult to define by normal reflection seismology.

CHAPTER 15

Enhanced Oil Recovery (EOR)

Historically, water flooding and gas injection have been referred to as secondary recovery techniques and other, more exotic, techniques have been referred to as tertiary techniques. Today, the term *enhanced oil recovery (EOR)* includes both secondary and tertiary recovery techniques.

The worldwide average recovery efficiency for primary oil is on the order of 33 percent. This means that, after primary recovery, two thirds of the oil is left in the ground, principally as residual oil in water wet reservoirs. Attempts to recover this oil include:

- Secondary recovery techniques
  - —Water flooding
  - —Gas injection
- Tertiary recovery techniques
  - —Chemical flooding processes:
    - Polymer flooding
    - Surfactant-polymer flooding
    - Caustic flooding
  - —Thermal recovery processes:
    - Steam flooding
    - Cyclic injection (huff and puff)
    - Steam drive
    - In-situ combustion

Hot water
Electromagnetic (microwaves)
—Miscible recovery processes:
Miscible hydrocarbon displacement
Carbon dioxide injection
Inert gas injection (nitrogen or flue gas)
—Microbial EOR

15.1 Some EOR Principles

The amount of oil that is recovered by water flooding and by other EOR injection techniques is a function of the following:

1. The amount of oil in place
2. *Volumetric sweep efficiency*—the percentage of oil that is contacted by the flood, which is a function of:
   (a) *Areal sweep efficiency* and
   (b) *Vertical sweep efficiency*
3. *Displacement efficiency*—the percentage of contacted oil that is moved or displaced.

15.1.1 Sweep efficiency

In making estimates of volumetric sweep efficiencies for reservoir simulations, geology must be considered. The geologist, through knowledge of facies patterns, must attempt to help the engineer to define flow units within the reservoir. Flow units (defined later this chapter) are three-dimensional rock bodies in which fluids are likely to behave similarly. They are not necessarily defined by facies, but are areas of high and low permeability that are likely to behave similarly under given flow conditions.

15.1.2 Mobility ratio

One of the most important factors in determining sweep efficiency is the mobility ratio between the two fluids. If the injected fluid is more mobile than the oil, the injected fluid is very likely to finger ahead through zones of high permeability and bypass much of the oil. *Fluid mobility* is a function of relative permeability for that fluid and the reciprocal of its viscosity. The mobility ratio is the ratio between the injected fluid divided by the displaced phase. If the ratio is less than 1, the injected fluid should not bypass the displaced fluid. For untreated water the mobility ratio between normal oil and water is greater than 1, and water has a strong tendency to bypass oil.

15.1.3 Displacement efficiencies

Methods that significantly help move oil through rock are:

1. Increase the mobility of the oil relative to the water by:
   (a) Increasing the viscosity of the water. This is the principle behind polymer flooding. Polymers added to water increase the viscosity of the water, and thus, they lower the mobility of the water.
   (b) Decreasing the viscosity of the oil. This is part of the principle behind steam flooding and in-situ combustion. Viscosity of oil decreases significantly as it is heated.
2. Help the oil move through the pore throats. This can be helped by:
   (a) Changing the interfacial tension at the pore throat
   (b) Changing the wetability characteristics of the rock
   (c) Changing the relative permeability of the fluids

Surfactants can play an important role in all three of the above, and when combined with polymers, they can make a very effective (though expensive) flood.

Another important means of getting oil through the pore throats is to dissolve the oil, in a more mobile solvent or dissolve a solvent in the oil, either of which makes the oil more mobile. Both *carbon dioxide floods* and *miscible hydrocarbon displacement* methods employ this principle.

15.2 Secondary Recovery Techniques

15.2.1 Water flooding

Water flooding is used on a routine basis throughout the world to maintain reservoir pressure and to push oil in front of a water front. Injected water is normally taken from the subsurface because surface waters or seawater commonly react with formation waters to cause undesirable precipitates or expansion of clay minerals. Whatever the origin of the water, its chemistry must be checked carefully against the chemistry of the formation fluids to make certain that the fluids are chemically compatible. It can be quite embarrassing to start injecting water only to find that the formation is now impermeable because of precipitates such as

$BaSO_4$ or other insoluble minerals caused by mixing of incompatible fluids. Such insoluble precipitates can virtually shut off permeability around injection wells.

Configuration of the reservoir, depth, and cost are important factors in determining which type of injection pattern to implement in a water flood. Typically, for thin, steeply dipping reservoirs (Fig. 15–1), an edge water drive is expected. Water injectors are placed near the original water level and producing wells are placed updip. As the oil-water contact moves updip and producing wells water out, the production wells are progressively converted to water injectors.

In such a configuration (Fig. 15–1), wells are expected to water out progressively from row 6 to row 5 to row 4, and row 3 if the sweep is good. Unfortunately, in poor and inhomogeneous reservoirs it is not uncommon for an updip well (such as in row 4) to water out first. This commonly happens when injected water fingers its way up through a zone of high permeability. This situation is undesirable because oil is commonly bypassed or left behind in isolated areas. In some cases, the geologist may be able to identify a facies pattern, such as a distributary channel within a delta front sand, where such fingering or channeling is likely to occur, and designs can sometimes be adjusted to accommodate for this occurrence. For thick reservoirs in relatively flat-lying rocks, where bottom water is present, or where no water is present, other configurations, such as those shown in Figure 15–2, are commonly used.

15.2.2 Gas injection

For gas reservoir associated with oil, it is of course undesirable (illegal in some cases) to blow down the gas cap before the oil is produced. Wells that produce from near the gas-oil contact commonly produce with a high gas-oil ratio (GOR) as gas is coned downward into the oil perforations. In order to maintain reservoir pressure, solution gas is commonly compressed and injected back into the gas cap. Alternatively, the compressed gas may be injected into the annulus, where it passes through gas lift mandrels into the tubing. This injected gas helps lighten the oil column in the tubing, which in turn helps lift the oil to the surface. Gas lift, like pumping, is a primary oil recovery technique.

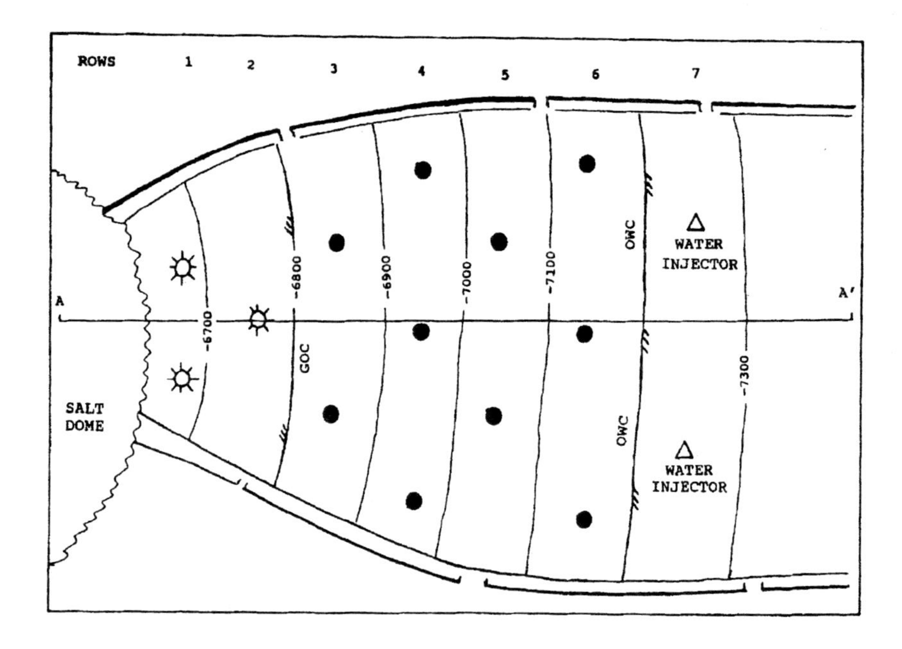

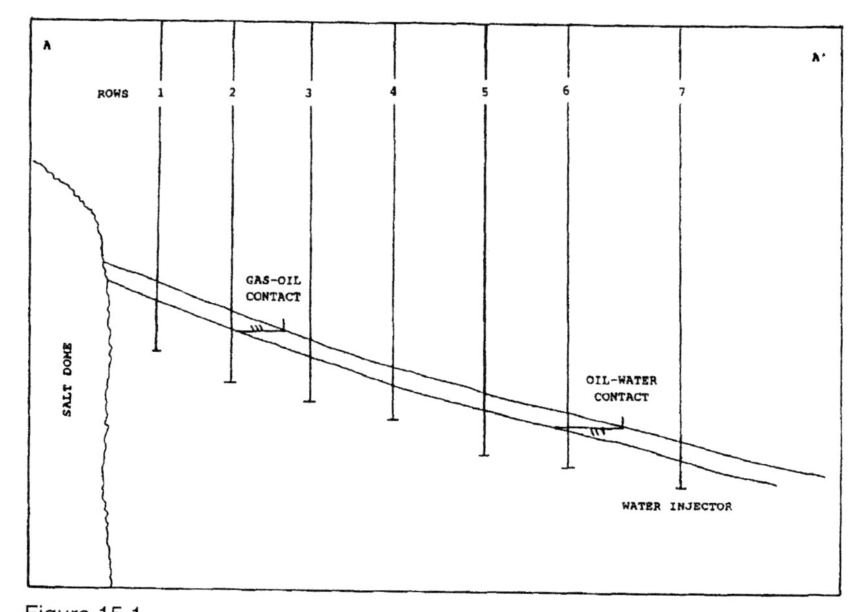

Figure 15.1
Map and cross section through highly dipping sediments showing edge water drive. Wells are expected to water out and be converted to water injectors successively from rows 6 to 5 to 4.

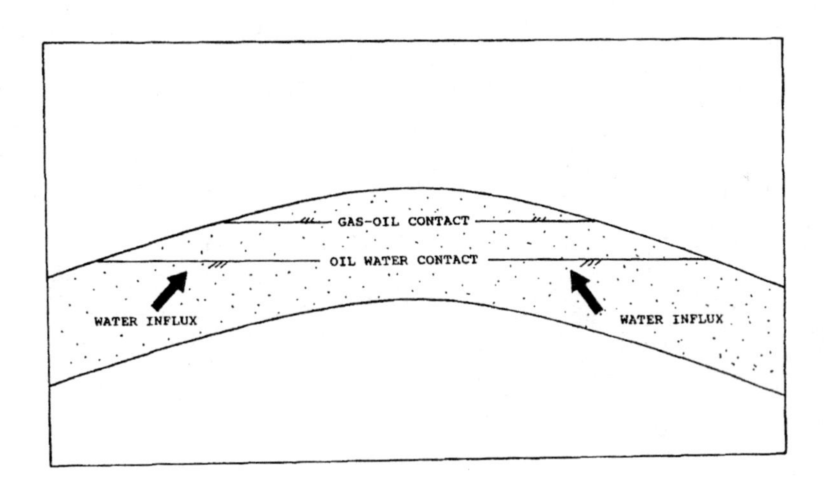

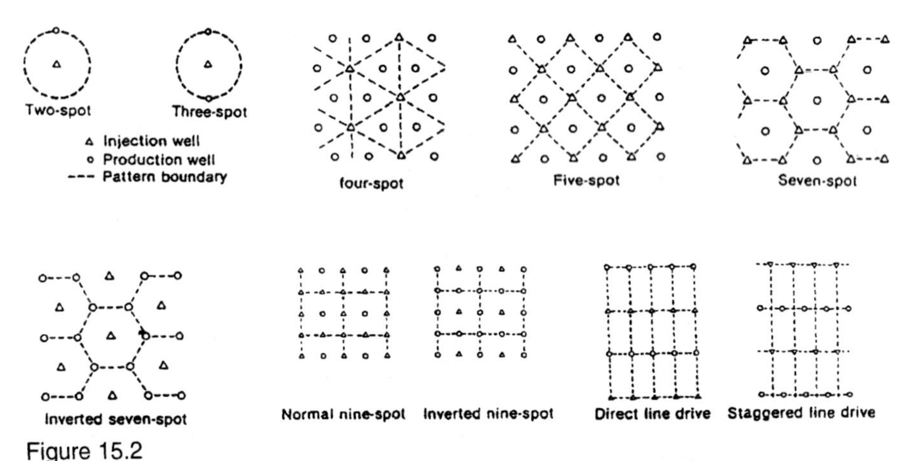

Figure 15.2
Cross section view of a bottom water drive reservoir and map view of typical well spot patterns for water flood injection design.

15.3 Tertiary Recovery Techniques

15.3.1 Chemical floods

Polymer flooding Polymers are commonly added to injection water to make the water more viscous, thus reducing the water's mobility. This tends to plug up the high-permeability zones which will normally improve sweep efficiency. Foam is another way to decrease the mobility of the displacing fluid and create a more uniform sweep efficiency.

Surfactants Oil and water are *immiscible,* meaning that they form emulsions in the subsurface that can be very difficult to break. Emulsions are caused by high surface tension between the two fluids. Surfactants such as soap or micelles are commonly added to polymer floods to break the emulsions and decrease interfacial tension between oil and water. *Caustic soda* is used to increase the pH of the reservoir, and, depending on the type of crude, react with the oil to create surfactants that help move the oil through the pore throats. Caustic flooding is normally applied to relatively acid crudes with high API gravity numbers.

15.3.2 Thermal recovery processes

Steam flooding consists of two principal types. In relatively small reservoirs or reservoirs that do not have good lateral permeability, a steam soak is commonly used. Steam is injected into the reservoir and allowed to soak into the formation for a period of time, commonly a week. The steam heats the oil and reduces its viscosity. The injection well is then produced for a period of time, also commonly a week. This procedure is referred to as a "huff and puff" process, as the well or wells are alternately used as injection wells and then as producing wells. In a conventional steam flood, steam is injected through injection wells and production occurs through production wells. Usually, this is a patterned flood such that the entire reservoir is swept. Surfactants are commonly added to help mobilize the oil. Steam floods are normally used at relatively shallow depths because, as pressure increases with depth, higher temperatures are required to keep water in vapor form.

In-situ combustion is a process whereby air or oxygen is injected into the formation where combustion of reservoir oil and gas can occur. The heat generated by the combustion creates a steam bank that drives the oil to producing wells.

Hot water injection is much like a water flood, except that the water is heated to reduce the viscosity of the oil.

Electromagnetic or heating by microwaves has been considered by a number of companies. Microwaves are very efficient at heating surfaces, but they do not penetrate more than a few centimeters past the surface. To date, no one has discovered a method to transmit the energy deep into the formation where it can do some real good in terms of mobilizing oil.

15.3.3 Miscible recovery processes

Miscible hydrocarbon displacement Oil and some natural gases, such as ethane, are *miscible,* meaning that the surface tension between the two phases is very low. Some natural gases tend to dissolve in oil, which can reduce the viscosity of the oil. Both of these processes can significantly help move relatively heavy and viscous oils through pore throats. Miscible floods tend to be quite expensive, but they can be attractive if the majority of the injected gas is ultimately recovered.

Carbon dioxide, nitrogen, and *flue gases* are the most commonly injected gases because they are cheap, they are often readily available as waste products, and they reduce the surface tension of the oil.

15.3.4 Microbial EOR

To date, no large-scale microbial projects have been attempted. There are two principal ideas behind microbial EOR. In the first method, microbes plus nutrients are injected into the formation. The microbes decompose the oil to produce detergents, CO_2, and new cells which either mechanically or chemically release oil from the reservoir pores.

In the second method, microbes and nutrients are injected into the reservoir, where they partially degrade the oil. Through this mechanism, the degraded oil and microbes block off areas of highest permeability such that further injection of other fluids causes the zones of lower permeability to be selectively flushed.

15.3.5 Comments on recovery

Finally, the injection of anything that increases a pressure gradient in the reservoir should help additional amounts of oil pop through the pore throats of the reservoir. Once oil is in discontinuous phase, it becomes very difficult to move. All of the above methods are enhanced by the injection of fluids in slugs such that the oil moves through the rock in zones of continuous phase oil.

15.4 Reservoir Modeling

When secondary and tertiary recovery techniques are being considered, reservoir engineers commonly develop computer programs that model each of the reservoirs. To work properly, such programs must consider all of the following:

1. Overall geometry of the reservoir:

(a) External shape
(b) Internal configuration of porosity and permeability:
- Horizontal permeability
- Vertical permeability
- Distribution of low permeability zones
- Distribution of fluid saturations
- Internal configurations of flow units
- Orientation and distribution of fractures

2. Location of injectors and producing wells
3. Pressure, volume, and temperature (PVT) data on the fluids to be produced
4. Type of recovery technique to be employed
5. Flow rates for injectors and producing wells

15.4.1 Reservoir characterization

Some reservoir modelers assume that the internal configuration of the reservoir is either isotropic and homogeneous, or so unpredictable that the geologist need not be consulted. The development geologist, based on correlation of well logs, petrographic data including porosity, and permeability data, plus regional data, almost always has some depositional and diagenetic models in mind for every reservoir. Rarely do those models assume an isotropic and homogeneous internal configuration. In order to model a reservoir properly the geologist and engineer must describe or characterize the reservoir in terms of all of the above characteristics. The geologist, in particular, must describe flow units.

15.4.2 Flow units

Flow units are reservoir units in the subsurface that are in hydrodynamic communication and have similar porosity and permeability characteristics. Commonly they are facies-dependent, but they need not be. For example, Figure 15–3 shows an example from the Pennsylvanian of the central U.S. where a prograding delta-front sandstone is overprinted by a meandering stream sequence. Although five general facies can be identified, only three flow units are important. From lower to upper, they are:

1. **Submarine delta front:** This facies is composed of either siltstone or thinly bedded alternating sandstone and mudstone. It is a transition zone between prodelta mudstone and delta front

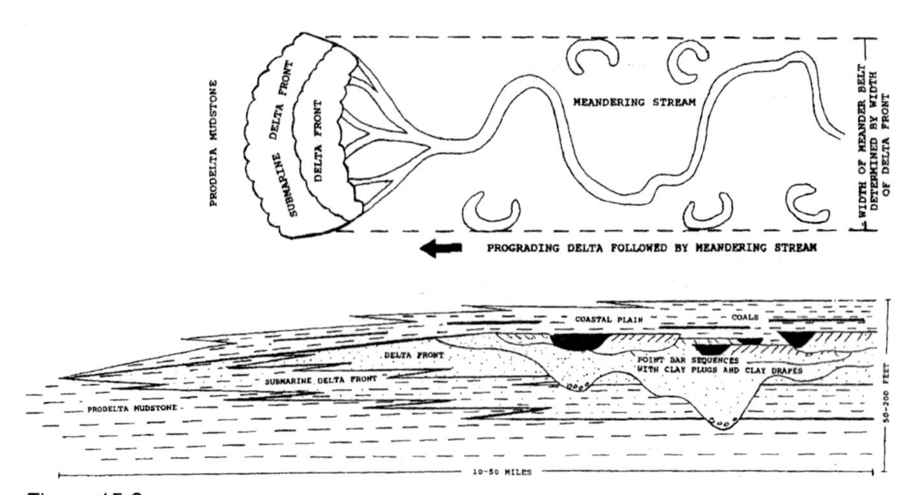

Figure 15.3
Map and cross section view of a typical Pennsylvanian river-dominated delta. The prograding delta front is overprinted by a meandering stream sequence. Flow units consist of lower, poor quality submarine delta-front siltstone or thin-bedded alternating sandstone and mudstone; high-quality delta-front sandstone and lower, point bar channel fill sequence; and upper point bar sequence which will have poor lateral continuity and highly partitioned reservoirs.

sandstone. It is composed of poor quality reservoir rock and will be very difficult to both produce and water flood.

2. **Delta-front sandstone and lower point bar sequence:** This is the best quality reservoir rock and, even though depositional environments are very different, their flow characteristics are similar for transmission of fluids.

3. **Upper point bar sequence:** Although oriented core samples of sandstone show high horizontal permeability, this area is likely to have clay plugs and clay drapes (see Chapter 12 for details) and have highly partitioned reservoirs. Horizontal permeability is likely to be highly restricted even though core data say otherwise.

15.4.3 Models

To model this reservoir as a homogeneous reservoir would certainly give incorrect results. The problem for the geologist is that, although it is known that the clay plugs and clay drapes are present, there is commonly no way of knowing exactly where they are.

A *deterministic* model is a model where the geologist is asked to make the best interpretation possible from the existing data. A deterministic model for the clay plugs may be possible if three-dimensional seismic imaging is available (Fig. 14–11), but more commonly the geologist knows that the permeability barriers are present in the reservoir, but has no way of knowing exactly where they are.

Newer *probabilistic* models (sometimes referred to as *stochastic* models) allow the geologist to assign statistical probabilities to the model. For example, in the point bar sequence example, the development geologist may be able to make the following probabilistic statements:

1. The lower part of the reservoir will have good horizontal permeability, and has an 80 percent chance of being in communication with a well located 1000 feet away.
2. A well drilled 1000 feet away has a 20 percent chance of being totally isolated from the first well by a clay plug that may cut down through the whole sequence.
3. The upper part of the reservoir, while having good horizontal permeability as determined from cores, is very likely to have extremely poor lateral continuity because of clay drapes. The upper part of the reservoir has a 90 percent chance that it will be shingled by impermeable clay drapes, and lateral continuity of the reservoir should not be expected to exceed 300 feet.
4. There is a 10 percent chance that the clay drapes have been destroyed by a chute cutoff. If this has happened, the upper part of the reservoir should have good lateral continuity.

These geologic models, coupled with the flow unit concept and reservoir simulation models, can create probabilistic (or statistical) models that are far superior to the older deterministic models. Not only are the models better, but they give full-range statistics that can be interpreted for errors.

The concept of reservoir characterization and definition of flow units within reservoirs are fundamental to all secondary recovery and EOR programs. Geologic input is one of the most important aspects of reservoir characterization, and if there is any place where the geologist, engineer, and statistician must work together, this is it.

CHAPTER 16

Some Concluding Remarks

Development geology is not a unique discipline. It is a hybrid, a mixture of many different fields, primarily geology, reservoir engineering, petrophysical engineering, geophysics, economics, and mostly common sense. There are no magical formulas that solve all or even very many of the development geologist's problems.

The principal charge of the development geologist is the business of making structure contour maps, isopachs, and reserve estimates. The best development geologists are people who can mentally visualize the traps and the reservoirs in three dimensions. Visualizing in three dimensions is an ability that is very difficult to teach. Some say that it can't be taught at all, and I have met individuals who simply were not able to visualize the traps and reservoirs in three dimensions. There is no substitute for an instinctive feeling as to whether a field looks right or not. This comes only from experience.

Many geologists don't particularly like it when they get an assignment in the production department. They feel that development geology is not as glamorous as exploration. Explorationists are supposed to have new and unique and sometimes bizarre ideas. That is their job. But development geologists are individuals who must have common sense. They must, absolutely, be realists. Development geologists are commonly doers rather than dreamers. They are people who like to work with other people to solve complex problems and get things done.

There is a certain unexplainable satisfaction that comes from working a field from beginning to end, and straightening out areas that are not right. When a development geologist is assigned an old field, there is a natural tendency to assume that the old geology is correct or that it is as correct as can be known, given the data. Sometimes the old geology is good. Sometimes it is not. It is extremely satisfying to take over an old field, particularly if the old field has areas that don't make sense, and rework the field into something that makes sense. It is like solving a three-dimensional puzzle. Instinct can guide the sense that the field is right, and it is very satisfying to make it right, particularly if additional wells can be drilled and new reserves can be discovered.

In onshore areas, it is not uncommon to put a field on line and let it flow for very long periods of time (as much as 50 years). In the offshore where production is from freestanding caissons or from platforms, fields must be designed to be drained within the expected life of the platform or caisson. Most platforms are designed to last 20 years, and reservoirs are designed to be drained within a 10- to 12-year timeframe. In the offshore, it is important that development be good *from the beginning,* because after the platform has rusted away, the economics of redevelopment are much less attractive.

16.1 Computers

One of the biggest changes that has occurred in the petroleum industry over the past 20 years involves computers, particularly the personal computer (PC). Computers have had a great impact in several particular areas that are important to the development geologist:

1. Word processing (this may affect clerical staff more than geologists)
2. Reservoir simulation (may affect engineers more than geologists)
3. Database management systems
4. Digitization of well logs
5. Computer contouring
6. Seismic analysis, particularly three-dimensional seismic imaging.

Word processing systems may not affect all development geologists, but they certainly help clerical staff and independent geologists. They have helped this author tremendously because I can't read my own writing, and I can't spell. My spell checker has taught me that I have misspelled certain words all my life. The PC has revolutionized report writing.

Reservoir engineers are capable of producing excellent reservoir simulation programs that could only be done using mainframe equipment ten years ago. Some excellent reservoir programs are now available for the PC.

Database management systems (such as dBase, Lotus 1-2-3, or Quattro) are presently revolutionizing how to keep track of sand tops, isopach intervals, and much more. The databases can be interfaced directly with contour and isopach programs. They can be interfaced with well-log data for direct retrieval or for direct interpretation of well-log data. Such data can be sorted and retrieved in any order or format. The programs are truly excellent, but they are no better than the data that is installed in the original database. There is also a substantial learning and frustration curve that goes with the learning of each database and computer system.

Digitization of well logs is a truly important new tool for the development geologist. Old, uncalibrated well logs that were plotted on a linear scale may now be digitized, plotted on a logarithmic scale, and calibrated against newer logs. This is extremely important for geologists who are working in fields where most of the logs are decades old. These old logs can also be compared against recently run logs to determine which zones have been drained and which have not.

There are now some truly excellent computer contouring packages on the market. Many of the packages are capable of making spectacular three-dimensional diagrams (Figs. 14–10 and 14–11). Experience with computer contouring packages has shown the following:

1. The less expensive the package, the less it is capable of doing. This is intentionally stated backwards because it does not necessarily follow that the more expensive the package, the more its capabilities.
2. Computer contouring packages do not handle faults, particularly dipping faults, well. Some systems handle faults well if the locations of the faults are well known, and can be input into the computer. But no computer begins to compare to the intuition that occurs in the human mind when it comes to sorting out fault patterns in complex fields.

3. Computer-generated maps are an effective tool for the development geologist, but they should not be used as the final product. Computers are excellent for creating mathematically derived, unbiased maps. They are not good for handling unusual situations or anything where geologic bias is important.
4. Computers can be used very effectively when combined with database systems for reconnaissance mapping.
5. Computers are a very effective tool for the explorationist, but they should not be used as a substitute for human perception. Geology is a soft science. Computers will never replace the intuitive or artistic part of geology.

16.2 The Future of Development Geology

I would like to reiterate the importance that development geology is going to play in the future of this nation and ultimately the world. At the present time, 71 percent of the world's wells are in the United States, and they account for less than 3 percent of the world's proved reserves. Those wells are all declining and the reserves are not being replaced by significant new discoveries. The United States presently imports approximately 7.5 million barrels of oil per day. At $20 per barrel this is an imbalance of payments of $150 million per day or approximately $220 per U.S. citizen per year. The United States has an energy problem that is not going to be solved by conventional exploration. Future domestic oil and gas reserve additions will be based heavily on the extraction of more oil from existing fields. Internationally, future oil and gas reserve additions will come from conventional exploration as well as from improved secondary and tertiary recovery techniques and the reevaluation of old areas by procedures that are primarily the domain of the development geologist.

References

Anderson, E. M., 1951, Dynamics of faulting: Edinburgh, Oliver and Boyd Publishers, 191 p.

Baker, Ron, 1984, A primer of oilwell drilling: Austin, Texas, Petroleum Extension Service, 94 p.

Barker, Colin, 1982, Organic geochemistry in petroleum exploration: American Association of Petroleum Geologists Course Notes 10, 157 p.

Barrs, D. L., A. B. Copland, and W. D. Ritchie, 1970, Geology of Middle Devonian reefs, Rainbow Area, Alberta, Canada, *in* M. T. Halbouty, ed., Geology of giant petroleum fields: AAPG Memoir 14, p. 19–49.

Bates, C. C., 1953, Rational theory of delta formation: AAPG Bulletin, v. 37, p. 2119–2162.

Berg, R. R., 1962, Mountain flank thrusting in Rocky Mountain foreland, Wyoming and Colorado: AAPG Bulletin, v. 46, p. 704–707.

Bernard, H. A., R. J. Leblanc, and C. F. Major, 1962, Recent and Pleistocene geology of Southeast Texas, *in* E. H. Rainwater and R. P. Zingula, eds., Geology of the Gulf Coast and central Texas and guidebook of excursions: Houston Geological Society, GSA annual meeting, 1962, 391 p.

Billings, Marland P., 1972, Structural geology, 3rd ed.: Englewood Cliffs, N. J., Prentice-Hall, 606 p.

Bourke, L., P. Lefiner, J. C. Trouiller, T. Fett, M. Grace, S. Luthi, O. Serra, and E. Standen, 1989, Using formation microscanner images: The Technical Review, v. 37, no. 1, p. 16–40.

Brown, Alistair R., 1988, Interpretation of three-dimensional seismic data; AAPG Memoir 42, 253 p.

Brown, G. B., D. L. Katz, R. C. A. Oberfell, 1948, Natural gasoline and volatile hydrocarbons: Tulsa, Natural Gasoline Association of America, 44 p.

Brown, L. F., and W. L. Fisher, 1980, Seismic stratigraphic interpretation and petroleum exploration; AAPG Continuing Education Course Note Series, no. 16, 181 p.

Budding, M. C., K. M. Eastwood, J. C. Herweijer, S. E. Livera, A. H. M. Paardekam, and J. M. M. Regtien, 1988, Probabilistic modelling of discontinuous reservoirs: Proceedings Indonesia Petroleum Association, 7th Annual Convention, p. 15–24.

Candido, A., and N. C. Wardlaw, 1985, Reservoir geology of the Carmopolis Oil Field, Brazil: Bulletin Canadian Petroleum Geology, v. 33, no. 4, p. 379–395.

Carr, N. L., R. Kobayashi, and D. B. Burrows, 1954, Viscosity of hydrocarbon gases under pressure: Transactions American Institute of Mining Engineers, v. 201, p. 270.

Choquette, P. W., and L. C. Pray, 1970, Geologic nomenclature and classification of porosity in sedimentary carbonates: AAPG Bulletin, v. 54, p. 207–250.

Clark, J. A., and M. T. Halbouty, 1972, The last boom: New York, Random House, 305 p.

________, 1985, Spindletop: New York, Random House, 306 p.

Craft, B. C., and M. F. Hawkins, 1959, Applied petroleum reservoir engineering: Englewood Cliffs, N. J., Prentice-Hall, 437 p.

De Buyl, M., 1989, Optimum field development with seismic reflection data: The Leading Edge, Society of Exploration Geophysicists, v. 8, no. 4, April 1989, p. 14–20.

DeGolyer & MacNaughton, 1985, Twentieth century petroleum statistics: Dallas, Texas, DeGolyer & MacNaughton, 65 p.

Dickey, Park A., 1979, Petroleum development geology: Tulsa, Petroleum Publishing Company, 398 p.

Dietrich, R. V., J. T. Dutro, Jr., and R. M. Foose, 1985, AGI data sheets: Falls Church, Virginia, American Geological Institute, 172 p.

Dott, Robert H., Jr., 1964, Wacke, greywacke, and matrix—what approach to immature sandstone classification?: Journal Sedimentary Petrology, v. 34, Fig. 3, p. 629.

Dunham, R. J., 1962, Classification of carbonate rocks according to depositional texture, *in* W. E. Ham, ed., Classification of carbonate rocks; AAPG Memoir 1, p. 108–121.

Ebanks, W. J., Jr., 1987, Flow unit concept—integrated approach to reservoir description for engineering projects (Abstract.): AAPG Bulletin, v. 71 no. 5, p. 551–552.

Economides, M. J., and K. G. Nolte, 1989, Reservoir stimulation: Englewood Cliffs, N. J., Prentice Hall, 338 p.

Farina, John, 1984, Geological applications of reservoir engineering tools: AAPG Continuing Education Course Notes Series, no. 28, 99 p.

Fisher, W. L., L. F. Brown, A. J. Scott, and J. H. McGowen, 1969, Delta systems in the exploration for oil and gas: Research Symposium, Austin, Texas, Bureau of Economic Geology, August 1969, 212 p.

Folk, Robert L., 1968, Petrology of sedimentary rocks: Austin, Texas, Hemphill's Bookstore, 185 p.

Folk, R. L., 1959, Practical petrographic classification of limestones: AAPG Bulletin, v. 43, p. 1–38.

Freidman, G. M., and J. E. Sanders, 1978, Principles of sedimentology: New York, John Wiley & Sons, 792 p.

Galloway, W. E., T. T. Ewing, C. M. Garret, N. Tyler, and D. G. Bebout, 1983, Atlas of major Texas oil reservoirs: Austin, Texas, Bureau of Economic Geology, 139 p.

Galloway, W. E., and D. K. Hobday, 1983, Terrigenous clastic depositional systems: New York, Springer-Verlag, 423 p.

Givens, W. W., 1987, A conductive rock matrix model for the analysis of low contrast resistivity formation: The Log Analyst, v. 28, no. 2, p. 138–151.

Gries, Robbie, 1983, Oil and gas prospecting beneath Precambrian of foreland thrust plates in Rocky Mountains: AAPG Bulletin, v. 67, p. 1–28.

Halbouty, M. T., ed., 1974, Geology of giant petroleum fields: AAPG Memoir 14, 575 p.

Halbouty, M. T., 1967, Salt Domes, Gulf Region, United States and Mexico; Houston, Gulf Publishing Company, 425 p.

Haldorsen, H. H., and L. W. Lake, 1984, A new approach to shale management in field-scale models: Journal Society Petroleum Engineering, August, 1984, p. 447–457.

Halstead, Philip H., and Edgar L. Berg, 1989, Oil field development optimization with seismic data: The Leading Edge, Society of Exploration Geophysicists, v. 8/7, July 1989, p. 15–22.

Handin, J., and R. V. Hager, Jr., 1957, Experimental deformation of sedimentary rocks under confining pressure: Tests at room temperature on dry samples: AAPG Bulletin, v. 41, p. 1–50.

Handin, J., and R. V. Hager, Jr., 1958, Experimental deformation of sedimentary rocks under confining pressure: Tests at high temperature: AAPG Bulletin, v. 42, p. 2892–2934.

Harris, D. G., 1975, The role of geology in reservoir simulation studies: Journal Petroleum Technology, May 1975, p. 625–632.

Harris, D. G., and C. H. Hewitt, 1977, Synergism in reservoir management—The geologic perspective: Journal Petroleum Technology, July 1977, p. 761–770.

Haugen, S. A., O. Lund, and L. A. Hoyland, 1988, Statfjord Field—Development strategy and reservoir management: Journal Petroleum Technology, July 1988, p. 863–73.

Hearn, C. L., W. J. Ebanks, Jr., R. S. Tye, and V. Ranganthan, 1984, Geological factors influencing reservoir performance of the Hartzog Draw Field, Wyoming: Journal Petroleum Technology, August 1984, p. 1335–1344.

Hemphill, C. R., R. I. Smith, and F. Szabo, 1970, Geology of Beaverhill Lake Reefs, Swan Hills area, Alberta, *in* M. T. Halbouty, ed., Geology of giant petroleum fields: AAPG Memoir 14, p. 50–90.

Hill, P. J., and G. V. Wood, 1980, Geology of the Forties Field, U. K. Continental Shelf, North Sea, *in* M. T. Halbouty, ed., Giant oil and gas fields of the decade: 1968–1978, AAPG Memoir 30, p. 81–94.

Hinch, H. H., 1980, The nature of shales and the dynamics of hydrocarbon expulsion in the Gulf Coast Tertiary section: AAPG Studies in Geology, no. 10, 18 p.

Hintze, L. F., 1982, Geologic highway map of Utah: Brigham Young University Geology Studies, Special publication #3. Provo, Utah, Brigham Young University Press.

Hull, C. E., and H. R. Warman, 1970, Asmari oil fields of Iran, *in* M. T. Halbouty, ed., Geology of giant petroleum fields: AAPG Memoir 14, p. 428–437.

Hunt, J. M., 1979, Petroleum geochemistry and geology: New York, W. H. Freeman, 617 p.

Jennings, J. B., 1987, Capillary pressure techniques: Application to Exploration and Development Geology: AAPG Bulletin, v. 71, no. 10, p. 1196–1209.

Jones, H. P., and R. G. Speers, 1976, Permo-Triassic reservoirs of Prudhoe Bay field, North slope, Alaska, *in* Jules Braunstein, ed., North American oil and gas fields, AAPG Memoir 24, p. 23–50.

Journel, A. G., and J. J. Gomez-Hernandez, 1989, Stochastic imaging of the Wilmington clastic sequence: Society of Petroleum Engineers, preprint 19857, 17 p.

Kamal, M. M., 1979, The use of pressure transients to describe reservoir heterogeneity: Journal Petroleum Technology, August 1979, p. 1060–70.

Koederitz, L. F., A. H. Harvey, and M. Honarpour, 1989, Introduction to petroleum reservoir analysis: Houston, Texas, Gulf Publishing Company, 250 p.
Krause, F. F., H. N. Collins, D. A. Nelson, S. D. Machemer, and P. R. French, 1987, Multiscale anatomy of a reservoir—geological characterization of Pembina-Cardium Pool, west-central Alberta, Canada: AAPG Bulletin, v. 71, no. 10, p. 1233–60.
Lake, Larry W., 1989, Enhanced oil recovery: Englewood Cliffs, N. J., Prentice-Hall, 550 p.
Lamb, Charles, F., 1980, Painter Reservoir Field: giant in the Wyoming thrust belt, *in* M. T. Halbouty, ed., Giant oil and gas fields of the decade: 1968–1978: AAPG Memoir 30, p. 281–288.
Landes, Kenneth K., 1970, Petroleum geology of the United States: New York, Wiley Interscience, 571 p.
Langston, E. P., J. A. Shirer, and D. E. Nelson, 1981, Innovative reservoir management—key to highly successful Jay/LEC waterflood: Journal of Petroleum Technology, May 1981, p. 783–791.
LeRoy, L. W., and D. O. LeRoy, 1977, Subsurface geology: Golden, Colorado, Colorado School of Mines, 940 p.
Levorsen, A. I., 1967, Geology of petroleum: New York, W. H. Freeman, 724 p.
Lucas, P. T., and J. M. Drexler, 1976, Altamont-Bluebell—a major, naturally fractured stratigraphic trap, Uinta Basin, Utah, *in* Jules Braunstein, ed., North American oil and gas fields: AAPG Memoir 24, p. 121–135.
Lucia, F. J., 1981, Petrophysical parameters estimated from visual description of carbonate rocks—a field classification of carbonate pore space: Journal Petroleum Technology, March 1981, p. 629–637.
Maher, C. E., 1980, Piper Oil Field, *in* M. T. Halbouty, ed., Giant oil and gas fields of the decade: 1968–1978: AAPG Memoir 30, p. 131–172.
Mason, John, F., and P. A. Dickey, eds., 1989, Oil field development techniques: Proceedings of the Daqing International Meeting, 1982: AAPG Studies in Geology, no. 28, 247 p.
Matthews, C. S., and D. G. Russell, 1967, Pressure buildup and flow tests in wells: Society of Petroleum Engineers Monograph, v. 1, Henry L. Doherty Series, p. 123.
Mayuga, M. N., 1970, Geology and development of California's Giant-Wilmington Field, *in* M. T. Halbouty, ed., Geology of giant petroleum fields: AAPG Memoir 14, p. 158–184.
McGregor, A. A., and C. A. Biggs, 1970, Bell Creek Field, Montana: a rich stratigraphic trap, *in* M. T. Halbouty, ed., Geology of giant petroleum fields: AAPG Memoir 14, p. 128–146.
Megill, Robert E., 1977, Risk Analysis: Tulsa, Petroleum Publishing Company, 199 p.
Megill, Robert E., 1979, Exploration economics: Tulsa, Petroleum Publishing Company, 180 p.
Morgridge, D. L., and W. B. Smith, Jr., 1972, Geology and discovery of Prudhoe Bay Field, eastern Arctic Slope, Alaska, *in* R. E. King, ed., Stratigraphic oil and gas fields—classification, exploration methods, and case histories: AAPG Memoir 16, p. 489–501.
Nelson, Ronald A., 1985, Geologic analysis of naturally fractured reservoirs: Houston, Gulf Publishing Company, 320 p.
Nelson, R. A., L. C. Lenox, and B. J. Ward, Jr., 1987, Oriented core: its use, error and uncertainty: AAPG Bulletin, v. 71 p. 357–367.
Newendorp, Paul D., 1975, Decision analysis for petroleum exploration: Tulsa, Petroleum Publishing Company, 668 p.
North, F. K., 1985, Petroleum geology: London, Allen & Unwin, 607 p.

Ottman, R. D., P. L. Keyes, and M. A. Ziegler, 1976, Jay Field, Florida—a Jurassic stratigraphic trap, *in* Jules Braunstein, ed., North American oil and gas fields: AAPG Memoir 24, p. 276–286.

Pippin, Lloyd, 1974, Panhandle-Hugoton Field, Texas-Oklahoma-Kansas—the first fifty years, *in* M. T. Halbouty, ed., Geology of giant petroleum fields: AAPG Memoir 14, p. 204–222.

Pullin, Norman, Larry Matthews, and Keith Hirsche, 1987, Techniques applied to obtain very high resolution 3-D seismic imaging at an Athabasca Tar Sands thermal pilot: The Leading Edge, Society of Exploration Geophysicists, v. 6/12, December 1988, p. 10–15.

Ranganathan, Vishnu, 1986, Petrography, diagenesis and facies controls on porosity in Shannon Sandstone, Hartzog Draw Field, Wyoming: AAPG Bulletin, v. 70, p. 56–69.

Richardson, Joseph G., 1989, Appraisal and development of Reservoirs: The Leading Edge, Society of Exploration Geophysicists, v. 8/2, February 1989, p. 42–44.

Robertson, James D., 1989, Reservoir management using 3-D seismic data: The Leading Edge, Society of Exploration Geophysicists, v. 8/2, February 1989, p. 25–31.

Roehl, P. O., and P. W. Choquette, 1985, eds., Carbonate petroleum reservoirs: New York, Springer-Verlag, 622 p.

Schlumberger, 1986, Schlumberger dipmeter interpretation fundamentals: Houston, Schlumberger, 76 p.

Schlumberger, 1989, Log interpretation principles/applications: Houston, Schlumberger Educational Services, 223 p.

Schramm, M. W., Jr., E. V. Dedman, and J. P. Lindsey, 1977, Practical stratigraphic modeling and interpretation, in Charles E. Payton, ed., Seismic stratigraphy—applications to hydrocarbon exploration: AAPG Memoir 26, p. 477–502.

Selley, Richard C., 1985, Elements of petroleum geology: New York, W. H. Freeman, 449 p.

Seni, S. J., and M. P. A. Jackson, 1983a, Evolution of salt structures, East Texas Diapir Province, Part 1: Sedimentary record of halokinesis: AAPG Bulletin, v. 67, p. 1219–1244.

Seni, S. J., and M. P. A. Jackson, 1983b, Evolution of salt structures, East Texas Diapir Province, Part 2; Pattern and rates of halokinesis: AAPG Bulletin, v. 67, p. 1245–1274.

Sneider, R. M., F. H. Richardson, D. D. Paynter, R. E. Eddy, and I. A. Wyant, 1977, Predicting reservoir rock geometry and continuity in Pennsylvanian reservoirs, Elk City Field, Oklahoma: Journal of Petroleum Technology, July 1977, p. 851–866.

Sneider, R. M., 1988, Reservoir description for exploration and development: what is needed and when?: AAPG Bulletin, v. 72, no. 12, p. 1522–23.

Sneider, R. M., W. Massell, R. Mathis, D. Loren, and P. Wichmann, 1990, Conveners and editors, The integration of geology, geophysics, petrophysics, and petroleum engineering in reservoir delineation, description and management: Proceedings of the First Archie Conference, Houston, Texas, October 1990, 441 p.

Standing, M. B., 1952, Volumetric and phase behavior of oil field Hydrocarbon Systems: New York, Van Nostrand Reinhold, 122 p.

________, and D. L. Katz, 1942, Density of natural gases: Transactions American Institute Mining Engineers, v. 146, p. 144.

Stearns, D. W., 1978, Faulting and forced folding in the Rocky Mountain Foreland, *in* V. Matthews III, ed., Laramide folding associated with basement block faulting in the Western United States: Geological Society of America Memoir 151, p. 1–36.

———, and M. Friedman, 1972, Reservoirs in fractured rocks: *in* Stratigraphic Oil and Gas Fields—Classification, Exploration Methods, and Case Histories: AAPG Memoir 10, p. 82–106.

Stewart, Robert, B., and James P. Disiena, 1989, The value of VSP in intepretation: The Leading Edge, Society of Exploration Geophysicists, v. 7/12, December 1989, p. 16–23.

Swanson, R. G., 1981, Sample examination manual: Methods in Exploration Series, Tulsa, American Association of Petroleum Geologists, 118 p.

Swift, D. J. P., D. B. Duane, and T. F. McKinney, 1973, Ridge and swale topography of the Middle Atlantic Bight, North America: Marine Geology, v. 15, p. 227–247.

Tearpock, Daniel, J., and Richard E. Bischke, 1991, Applied subsurface geological mapping: Englewood Cliffs, N. J., Prentice-Hall, 648 p.

Tillman, R. W., and D. W. Jordan, 1987, Sedimentology and subsurface geology of deltaic facies, Admire 650' Sandstone, El Dorado Field, Kansas: Society of Economic Paleontologists and Mineralogists Special Publication 40, p. 221–291.

Tissot, B. P., and D. H. Welte, 1978, Petroleum formation and occurrence: New York, Springer-Verlag, 538 p.

Tyler, N., and W. A. Ambrose, 1985, Facies architecture and production characteristics of strandplain reservoirs in the Frio Formation, Texas: University Texas Bureau of Economic Geology, report of investigation no. 146, 42 p.

Vest, E. L., Jr., 1970, Oil fields of Pennsylvanian-Permian Horseshoe Atoll, West Texas, *in* M. T. Halbouty, ed., Geology of giant petroleum fields: AAPG Memoir 14, p. 185–203.

Viniegra, O., and C. Castillo-Tejero, 1970, Golden Lane Fields, Veracruz, Mexico, *in* M. T. Halbouty, ed., Geology of giant petroleum fields: AAPG Memoir 14, p. 309–325.

Walker, R. G., and D. J. Cant, 1984, Sandy fluvial systems, *in* Roger G. Walker, ed., Facies Models: Toronto, Geoscience Canada, p. 71–90.

Wardlaw, N. C., 1983, Rocks, pores, and enhanced oil recovery—a geological challenge: AAPG Bulletin, v. 67, no. 11, p. 2157.

Wardlaw, N. C., and J. P. Cassan, 1978, Estimation of recovery efficiency by visual observation of pore systems in reservoir rocks: Bulletin Canadian Petroleum Geology, v. 26, p. 572–585.

Wasson, Theron, 1948, Creole Field, Gulf of Mexico, coast of Louisiana, *in* J. V. Howell, ed., Structure of typical American oil fields: AAPG Bulletin, v. 3, p. 281–298.

Weber, K. J., 1982, Influence of common sedimentary structures on fluid flow in reservoir models: Journal Petroleum Technology, March 1982, p. 665–672.

Index

C

F

G

H

M

N

O

T